Houghton
Mifflin
Harcourt

GO MATH!
FLORIDA

D1399724

GO MATH! FLORIDA

Houghton Mifflin Harcourt

Printed in the U.S.A.

ISBN 978-0-544-50082-2

2 3 4 5 6 7 8 9 10 0928 23 22 21 20 19 18 17 16

4500585002   ^ B C D E F G

Dear Students and Families,

Welcome to **Go Math!**, Grade 4! In this exciting mathematics program, there are hands-on activities to do and real-world problems to solve. Best of all, you will write your ideas and answers right in your book. In **Go Math!**, writing and drawing on the pages helps you think deeply about what you are learning, and you will really understand math!

By the way, all of the pages in your **Go Math!** book are made using recycled paper. We wanted you to know that you can Go Green with **Go Math!**

Sincerely,

The Authors

Made in the United States
Text printed on 100% recycled paper

GO MATH!

Authors

Juli K. Dixon
Professor of Mathematics Education
University of Central Florida
Orlando, Florida

Matt Larson
Curriculum Specialist for Mathematics
Lincoln Public Schools
Lincoln, Nebraska

Miriam A. Leiva
Founding President, TODOS:
    Mathematics for All
Distinguished Professor
    of Mathematics Emerita
University of North Carolina Charlotte
Charlotte, North Carolina

Thomasenia Lott Adams
Professor of Mathematics Education
University of Florida
Gainesville, Florida

Place Value and Operations with Whole Numbers

Developing understanding and fluency with multi-digit multiplication, and developing understanding of dividing to find quotients involving multi-digit dividends

1 Place Value, Addition, and Subtraction to One Million   3

Domain Number and Operations in Base Ten

DIGITAL PATH
Go online! Your math lessons are interactive. Use *i*Tools, Animated Math Models, the Multimedia *e*Glossary, and more.

Look for these:

Project Food in Space

REAL WORLD

H.O.T.
Higher Order Thinking

Connect to Science
p. 20

GO MATH! FLORIDA

Use every day
For Standards Practice.

© Houghton Mifflin Harcourt Publishing Company

4 Divide by 1-Digit Numbers 135

Domains Operations and Algebraic Thinking
Number and Operations in Base Ten

Look for these:

REAL WORLD

H.O.T.
Higher Order Thinking

Connect to Science

p. 152

Connect to Reading

p. 156

GO MATH! FLORIDA

Use every day
For Standards Practice.

5 Factors, Multiples, and Patterns 191

Domain Operations and Algebraic Thinking

Look for these:

REAL WORLD

H.O.T.
Higher Order Thinking

Connect to Social Studies

p. 214

GO MATH! FLORIDA

Use every day
For Standards Practice.

viii

© Houghton Mifflin Harcourt Publishing Company

Fractions and Decimals

Developing an understanding of fraction equivalence, addition and subtraction of fractions with like denominators, and multiplication of fractions by whole numbers

6 Fraction Equivalence and Comparison   225

Domain Number and Operations–Fractions

DIGITAL PATH
Go online! Your math lessons are interactive. Use *i*Tools, Animated Math Models, the Multimedia *e*Glossary, and more.

7 Add and Subtract Fractions   265

Domain Number and Operations–Fractions

Look for these:

Project Building Custom Guitars

REAL WORLD

H.O.T.
Higher Order Thinking

Connect to Art

p. 278

Use every day
For Standards Practice.

Look for these:

REAL WORLD

H.O.T.
Higher Order Thinking

Connect to Science

p. 354

Use every day
For Standards Practice.

Geometry, Measurement, and Data

Understanding that geometric figures can be analyzed and classified based on their properties, such as having parallel sides, perpendicular sides, particular angle measures, and symmetry

DIGITAL PATH
Go online! Your math lessons are interactive. Use *i*Tools, Animated Math Models, the Multimedia *e*Glossary, and more.

Look for these:

Project Landscape Architects

REAL WORLD

H.O.T.
Higher Order Thinking

Connect to Science
pp. 384, 428

Connect to Art
p. 396

GO MATH! FLORIDA

Use every day
For Standards Practice.

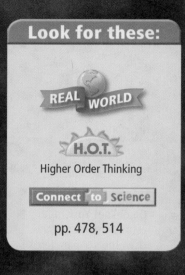

Look for these:

REAL WORLD

H.O.T.
Higher Order Thinking

Connect to Science

pp. 478, 514

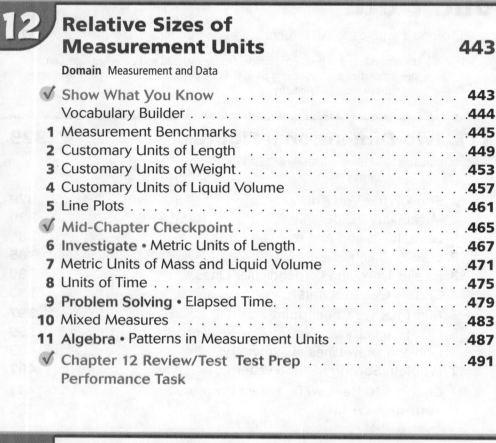

Use every day
For Standards Practice.

© Houghton Mifflin Harcourt Publishing Company

Place Value and Operations with Whole Numbers

Developing understanding and fluency with multi-digit multiplication, and developing understanding of dividing to find quotients involving multi-digit dividends

Space Shuttle launching from Kennedy Space Center ▶

Project

Food in Space

The United States is planning a manned mission to Mars. The crew must take all of its food along on the journey, because there is no food available on Mars.

Get Started

Work with a partner. You are in charge of planning the amount of food needed for the Mars mission. Decide how much food will be needed for the entire trip. Use the Important Facts to help you plan. **Explain** your thinking.

_____

_____

_____

_____

Important Facts

- Length of trip to Mars: 6 months
- Length of stay on Mars: 6 months
- Length of return trip to Earth: 6 months
- Number of astronauts: 6
- 2 cups of water weigh 1 pound.
- 1 month = 30 days (on average).
- Each astronaut needs 10 cups of water and 4 pounds of food each day.

Completed by _____

Show What You Know

Check your understanding of important skills.

Name _____

▶ **Tens and Ones** Write the missing numbers.

1. 27 = _____ tens _____ ones

2. 93 = _____ tens _____ ones

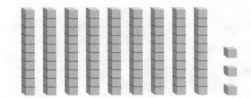

▶ **Regroup Hundreds as Tens** Regroup. Write the missing numbers.

3. 5 hundreds 4 tens = _____ tens

4. 8 hundreds 9 tens = _____ tens

▶ **Two-Digit Addition and Subtraction** Add or subtract.

5.   27
      + 34

6.   95
      + 46

7.   84
      − 27

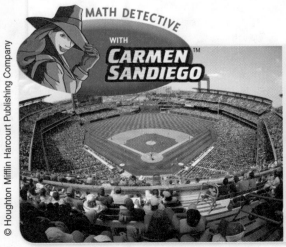

MATH DETECTIVE WITH **CARMEN SANDIEGO**™

The home stadium of the Philadelphia Phillies is a large baseball park in Philadelphia, PA. Be a Math Detective. Use the following clues to find the stadium's maximum capacity.

- The 5-digit number has a 4 in the greatest place-value position and a 1 in the least place-value position.
- The digit in the thousands place has a value of 3,000.
- The digit in the hundreds place is twice the digit in the thousands place.
- There is a 5 in the tens place.

Vocabulary Builder

▶ **Visualize It** •

Write the review words with a ✓ on the Word Line, from greatest to least place value.

Review Words

✓ hundreds

inverse operations

✓ ones

✓ tens

✓ ten thousands

✓ thousands

Preview Words

estimate

expanded form

period

round

standard form

word form

Place Value

greatest _____

_____

_____

_____

least _____

▶ **Understand Vocabulary** • • • • • • • • • • • • • • • • • •

Read the definition. Which word does it describe?

1. To replace a number with another number that tells about
   how many or how much _____

2. A way to write numbers by showing the value of each digit
   _____

3. A number close to an exact amount _____

4. Each group of three digits separated by commas in a
   multi-digit number _____

5. A way to write numbers by using the digits 0–9, with each
   digit having a place value _____

Name _____

Model Place Value Relationships

Essential Question How can you describe the value of a digit?

🔑 UNLOCK the Problem

🔑 Activity Build numbers through 10,000.

Materials ■ base-ten blocks

1	10	100	1,000	10,000

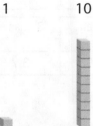 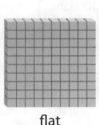 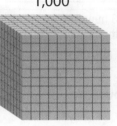

?

cube	long	flat	cube	_____
1	10 ones	_____ tens	_____ hundreds	_____ thousands

A small cube represents 1.

_____ small cubes make a long. The long represents _____.

_____ longs make a flat. The flat represents _____.

_____ flats make a large cube. The large cube represents _____.

Math Talk   MATHEMATICAL PRACTICES
Explain how you can use ten thousands longs to model 100,000.

1. **Describe** the pattern in the shapes of the models. What will be the shape of the model for 10,000?

   _____

   _____

   _____

2. **Describe** the pattern you see in the sizes of the models. How will the size of the model for 100,000 compare to the size of the model for 10,000?

   _____

   _____

   _____

Value of a Digit The value of a digit depends on its place-value position in the number. A place-value chart can help you understand the value of each digit in a number. The value of each place is 10 times the value of the place to the right.

 Write 894,613 in the chart. Find the value of the digit 9.

MILLIONS			THOUSANDS			ONES		
Hundreds	Tens	Ones	Hundreds	Tens	Ones	Hundreds	Tens	Ones
			8 hundred thousands	9 ten thousands	4 thousands	6 hundreds	1 ten	3 ones
			800,000	90,000	4,000	600	10	3

The value of the digit 9 is 9 ten thousands, or _____ .

 Compare the values of the underlined digits.

2,3_0_4   16,1_3_5

Math Talk

MATHEMATICAL PRACTICES

Explain how you can compare the values of the digits without drawing a model.

STEP 1 Find the value of 3 in 2,304.

Show 2,304 in a place-value chart.

THOUSANDS			ONES		
Hundreds	Tens	Ones	Hundreds	Tens	Ones

Think: The value of the digit 3 is _____ .

Model the value of the digit 3.

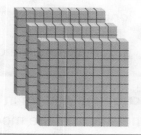

STEP 2 Find the value of 3 in 16,135.

Show 16,135 in a place-value chart.

THOUSANDS			ONES		
Hundreds	Tens	Ones	Hundreds	Tens	Ones

Think: The value of the digit 3 is _____ .

Model the value of the digit 3.

Each hundred is 10 times as many as 10, so 3 hundreds is ten times as many as 3 tens.

So, the value of 3 in 2,304 is _____ times the value of 3 in 16,135.

6

Name _____

Share and Show . . . .

1. Complete the table below.

Number	1,000,000	100,000	10,000	1,000	100	10	1
Model	?	?	?				
Shape				cube	flat	long	cube
Group				10 hundreds	10 tens	10 ones	1 one

Find the value of the underlined digit.

2. <u>7</u>03,890

3. 63,5<u>4</u>0

4. 1<u>8</u>2,034

✓ **5.** 34<u>5</u>,890

_____ _____ _____ _____

Compare the values of the underlined digits.

6. <u>2</u>,000 and <u>2</u>00

The value of 2 in _____ is _____

times the value of 2 in _____ .

✓ **7.** <u>4</u>0 and <u>4</u>00

The value of 4 in _____ is _____

times the value of 4 in _____ .

On Your Own . . . .

Find the value of the underlined digit.

8. 2<u>3</u>0,001

9. 803,0<u>4</u>0

10. 46,84<u>2</u>

11. <u>9</u>80,650

_____ _____ _____ _____

Compare the values of the underlined digits.

12. 6<u>7</u>,908 and <u>7</u>6,908

The value of 7 in _____

is _____ times the value of 7

in _____ .

13. 546,<u>3</u>00 and <u>3</u>,456

The value of 3 in _____

is _____ times the value of 3

in _____ .

Problem Solving REAL WORLD

Use the table for 14–15.

14. What is the value of the digit 7 in the population of Memphis?

_____

15. Which city's population has a 4 in the hundred thousands place?

_____

16. H.O.T. How many models of 100 do you need to model 3,200? Explain.

_____

_____

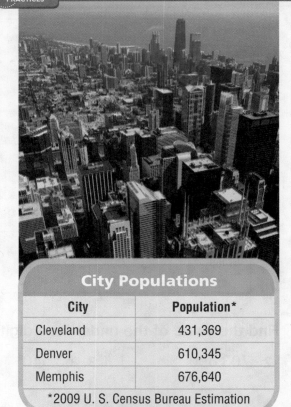

City Populations	
City	**Population***
Cleveland	431,369
Denver	610,345
Memphis	676,640
*2009 U. S. Census Bureau Estimation	

17. **Write Math** ▸ Sid wrote 541,309 on his paper. Using numbers and words, **explain** how the number would change if he switched the digits in the hundred thousands and tens places.

_____

_____

_____

_____

SHOW YOUR WORK

18. **Test Prep** There are 686,147 books at the Greenville Library. What is the value of the digit 8 in this number?

Ⓐ   80

Ⓑ 8,000

Ⓒ   80,000

Ⓓ 800,000

Name _____

Read and Write Numbers

Essential Question How can you read and write numbers through hundred thousands?

 **UNLOCK the Problem** REAL WORLD

The International Space Station uses 262,400 solar cells to convert sunlight to electricity.

Write 262,400 in standard form, word form, and expanded form.

 Use a place-value chart.

Each group of three digits separated by a comma is called a **period**. Each period has hundreds, tens, and ones. The greatest place-value position in the thousands period is hundred thousands.

Write 262,400 in the place-value chart below.

PERIOD   PERIOD

THOUSANDS			ONES		
Hundreds	Tens	Ones	Hundreds	Tens	Ones

The number 262,400 has two periods, thousands and ones.

Standard Form: 262,400

Word Form: two hundred sixty-two thousand, four hundred

Expanded Form: 200,000 + 60,000 + 2,000 + 400

Math Talk MATHEMATICAL PRACTICES
Which digit has the greatest value in 262,400? Explain.

Try This! Use place value to read and write numbers.

A **Standard Form:** _____

Word Form: ninety-two thousand, one hundred seventy

Expanded Form:

90,000 + 2,000 + _____ + 70

B **Standard Form:** 200,007

Word Form:
two hundred _____ , _____

Expanded Form:

_____ + 7

Share and Show 🖊️MATH BOARD ....

1. How can you use place value and period names to read and write 324,904 in word form?

_____

_____

_____

_____

Read and write the number in two other forms.

☑ 2. four hundred eight thousand, seventeen

_____

_____

☑ 3. 65,058

_____

_____

> **Math Talk** MATHEMATICAL PRACTICES
> Explain how you can use the expanded form of a number to write the number in standard form.

On Your Own ....

Read and write the number in two other forms.

4. five hundred eight thousand

_____

_____

5. forty thousand, six hundred nineteen

_____

_____

6. 570,020

_____

_____

7. 400,000 + 60,000 + 5,000 + 100

_____

_____

Use the number 145,973.

8. Write the name of the period that has the digits 145.

_____

9. Write the name of the period that has the digits 973.

_____

10. Write the digit in the ten thousands place.

_____

11. Write the value of the digit 1.

_____

Name _____

 Find the sum. Then write the answer in standard form.

12. 5 thousands 2 tens 4 ones
+ 4 thousands 3 hundreds 2 ones

_____

_____

13. 6 thousands 5 hundreds
+ 1 thousand 3 hundreds 4 tens

_____

_____

14. 4 ten thousands + 3 ten thousands
4 hundreds 8 tens

_____

_____

15. 4 ten thousands 3 ones + 1 ten thousand
9 hundreds 5 ones

_____

_____

Problem Solving REAL WORLD

Use the table for 16–17.

16. Which city has a population of two hundred fifty-five thousand, one hundred twenty-four?

_____

17. Write the population of Raleigh in expanded form and word form.

_____

_____

18. **H.O.T.** **What's the Error?** Sophia said that the expanded form for 605,970 is 600,000 + 50,000 + 900 + 70. **Describe** Sophia's error and give the correct answer.

_____

_____

_____

Major Cities in North Carolina	
City	Population*
Durham	229,171
Greensboro	255,124
Raleigh	405,612

*U.S. Census Bureau 2008 Estimated Population

▲ NC General Assembly/Legislative Building, Raleigh, North Carolina

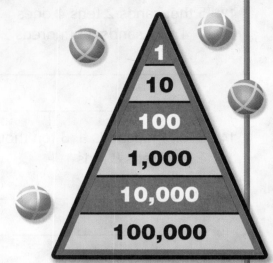
UNLOCK the Problem • REAL WORLD

19. Mark tossed six balls while playing a number game. Three balls landed in one section, and three balls landed in another section. His score is greater than one hundred thousand. What could his score be?

a. What do you know? _____

_____

_____

b. How can you use what you know about place value

to find what Mark's score could be? _____

_____

_____

_____

_____

_____

c. Draw a diagram to show one way to solve the problem.

d. Complete the sentences.

Three balls could have landed in the

_____ section.

Three balls could have landed in the

_____ section.

Mark's score could be _____

_____.

20. There are 2,750 sheep on a farm. Write the number of sheep in word form and expanded form.

_____

_____

21. Test Prep The new football stadium was filled to capacity with 105,840 fans. What is the value of the digit 5 in 105,840?

Ⓐ 500   Ⓒ 50,000

Ⓑ 5,000   Ⓓ 500,000

FOR MORE PRACTICE:
Standards Practice Book, pp. P5–P6

Name _____

Compare and Order Numbers

Essential Question How can you compare and order numbers?

🔑 UNLOCK the Problem REAL WORLD

Grand Canyon National Park in Arizona had 651,028 visitors in July 2008 and 665,188 visitors in July 2009. In which year did the park have more visitors during the month of July?

- How many visitors were there in July 2008?

  _____

- How many visitors were there in July 2009?

  _____

🔒 Example 1 Use a place-value chart.

You can use a place-value chart to line up the digits by place value. Line up the ones with the ones, the tens with the tens, and so on. Compare 651,028 and 665,188.

Write 651,028 and 665,188 in the place-value chart below.

THOUSANDS			ONES		
Hundreds	Tens	Ones	Hundreds	Tens	Ones

Start at the left. Compare the digits in each place-value position until the digits differ.

STEP 1 Compare the hundred thousands.

651,028

665,188

6 hundred thousands ◯ 6 hundred thousands
└ Write <, >, or =.

The digits in the hundred thousands place are the same.

STEP 2 Compare the ten thousands.

651,028

665,188

5 ten thousands ◯ 6 ten thousands
└ Write <, >, or =.

5 ten thousands is less than 6 ten thousands so, 651,028 < 665,188.

Since 651,028 < 665,188, there were more visitors in July 2009 than in July 2008.

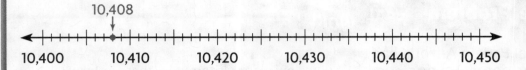

Example 2 Use a number line to order 10,408; 10,433; and 10,416 from least to greatest.

Locate and label each point on the number line. The first one is done for you.

10,408

10,400   10,410   10,420   10,430   10,440   10,450

Think: Numbers to the left are closer to 0.

So, the numbers from least to greatest are 10,408; 10,416; and 10,433. 10,408 < 10,416 < 10,433

Share and Show

1. Compare 15,327 and 15,341.
   Write <, >, or =. Use the number line to help.

15,300   15,310   15,320   15,330   15,340   15,350   15,360

15,327 ◯ 15,341

Compare. Write <, >, or =.

2. $631,328 ◯ $640,009

3. 56,991 ◯ 52,880

4. 708,561 ◯ 629,672

5. 143,062 ◯ 98,643

Math Talk MATHEMATICAL PRACTICES
Explain how you ordered the numbers from greatest to least in Exercise 6.

Order from greatest to least.

6. 20,650; 21,150; 20,890

_____

Round Numbers

Essential Question How can you round numbers?

🔑 UNLOCK the Problem REAL WORLD

During May 2008, the Mount Rushmore National Monument in South Dakota welcomed 138,202 visitors. A website reported that about 1 hundred thousand people visited the park during that month. Was the estimate reasonable?

> • Underline what you are asked to find.
> • Circle the information you will use.

An **estimate** tells you about how many or about how much. It is close to an exact amount. You can **round** a number to find an estimate.

🔓 One Way Use a number line.

To round a number to the nearest hundred thousand, find the hundred thousands it is between.

_____ < 138,202 < _____

Use a number line to see which hundred thousand 138,202 is closest to.

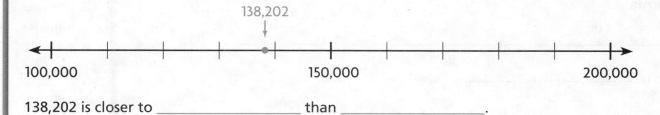

138,202 is closer to _____ than _____.

So, 1 hundred thousand is a reasonable estimate for 138,202.

Math Talk Is 155,000 closer to 100,000 or 200,000? **Explain.**

MATHEMATICAL PRACTICES

1. What number is halfway between 100,000 and 200,000?

_____

2. How does knowing where the halfway point is help you find which hundred thousand 138,202 is closest to? **Explain.**

_____

_____

_____

🔒 Another Way Use place value.

Mount Rushmore is located 5,725 feet above sea level. About how high is Mount Rushmore above sea level, to the nearest thousand feet?

To round a number to the nearest thousand, find the thousands it is between.

_____ < 5,725 < _____

Look at the digit in the place-value position to the right.

5,725
↑

Think: The digit in the hundreds place is 7.
So, 5,725 is closer to 6,000 than 5,000.

So, Mount Rushmore is about _____ feet above sea level.

MATHEMATICAL PRACTICES

Math Talk **Explain** how you know that 5,700 is closer to 6,000 than to 5,000.

3. What number is halfway between 70,000 and 80,000?

_____

4. What is 75,000 rounded to the nearest ten thousand? **Explain.**

_____

_____

_____

Math Idea
When a number is exactly half way between two rounding numbers, round to the greater number.

Try This! Round to the place value of the underlined digit.

A 6̲4,999

_____

C 301̲,587

_____

B 8̲50,000

_____

D 10̲,832

_____

Name _____

 Mid-Chapter Checkpoint

▶ **Check Vocabulary**

Choose the best term from the box.

1. The _____ of 23,850 is 20,000 + 3,000 + 800 + 50. (p. 9)

2. You can _____ to find *about* how much or how many. (p. 17)

3. In 192,860 the digits 1, 9, and 2 are in the same

   _____. (p. 9)

▶ **Concepts and Skills**

Find the value of the underlined digit.

4. 3<u>8</u>0,671   5. 10,6<u>9</u>8   6. <u>6</u>50,234

   _____   _____   _____

Write the number in two other forms.

7. 293,805   8. 300,000 + 5,000 + 20 + 6

   _____   _____

   _____   _____

   _____   _____

Compare. Write <, >, or =.

9. 457,380 ◯ 458,590   10. 390,040 ◯ 39,040   11. 11,809 ◯ 11,980

Round to the place of the underlined digit.

12. <u>1</u>40,250   13. 10,<u>4</u>50   14. 12<u>6</u>,234

   _____   _____   _____

© Houghton Mifflin Harcourt Publishing Company

Fill the bubble in completely to show your answer.

15. Last year, three hundred twenty-three thousand people visited the museum. What is this number in standard form?

Ⓐ 323,000

Ⓑ 323,300

Ⓒ 232,300

Ⓓ 232,000

16. Which number, rounded to the nearest hundred, is zero?

Ⓐ 94

Ⓑ 68

Ⓒ 52

Ⓓ 31

17. What is the highest volcano in the Cascade Range?

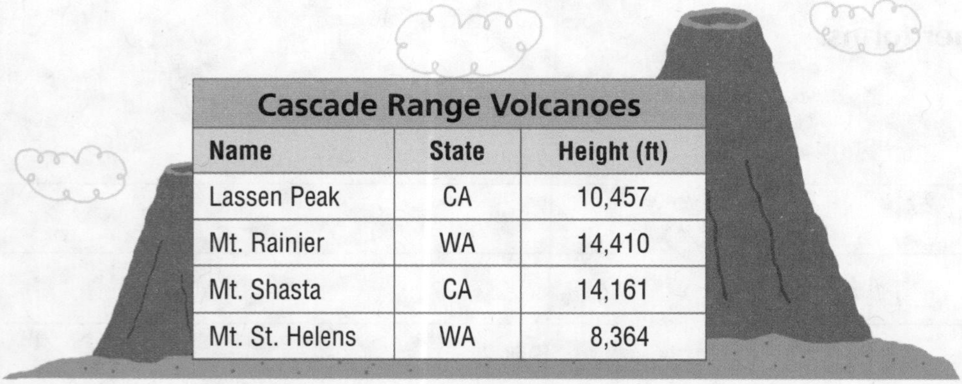

Cascade Range Volcanoes		
Name	**State**	**Height (ft)**
Lassen Peak	CA	10,457
Mt. Rainier	WA	14,410
Mt. Shasta	CA	14,161
Mt. St. Helens	WA	8,364

Ⓐ Lassen Peak

Ⓑ Mt. Rainier

Ⓒ Mt. Shasta

Ⓓ Mt. St. Helens

Name _____

Rename Numbers

Essential Question How can you rename a whole number?

Investigate

Materials ■ base-ten blocks

You can regroup numbers to rename them.

A. Use large cubes and flats to model 1,200. Draw a quick picture to record your model.

The model shows _____ large cube and _____ flats.

Another name for 1,200 is _____ thousand _____ hundreds.

B. Use only flats to model 1,200.
Draw a quick picture to record your model.

The model shows _____ flats.

Another name for 1,200 is _____ hundreds.

Draw Conclusions . . . .

1. How is the number of large cubes and flats in the first model related to the number of flats in the second model?

_____

_____

2. Can you model 1,200 using only longs? **Explain**.

_____

_____

3. You renamed 1,200 as hundreds. How can you rename 1,200 as tens? **Explain**.

_____

_____

4. **H.O.T.** **Apply** What would the models in Step A and Step B look like for 5,200? How can you rename 5,200 as hundreds?

_____

_____

_____

Make Connections ....

You can also use a place-value chart to help rename numbers.

THOUSANDS			ONES		
Hundreds	Tens	Ones	Hundreds	Tens	Ones
5	0	0,	0	0	0

└──────┘ 5 hundred thousands
└────────────┘ 50 ten thousands
└──────────────────┘ 500 thousands
└────────────────────────┘ 5,000 hundreds
└──────────────────────────────┘ 50,000 tens
└────────────────────────────────────┘ 500,000 ones

Write 32 hundreds on the place-value chart below. What is 32 hundreds written in standard form?

THOUSANDS			ONES		
Hundreds	Tens	Ones	Hundreds	Tens	Ones

└──────────────┘ 32 hundreds

32 hundreds written in standard form is _____.

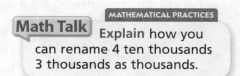
Math Talk MATHEMATICAL PRACTICES
Explain how you can rename 4 ten thousands 3 thousands as thousands.

Name _____

Share and Show

Rename the number. Draw a quick picture to help.

1. 150

_____ tens

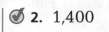 **2.** 1,400

_____ hundreds

3. 2 thousands 3 hundreds

_____ hundreds

4. 13 hundreds

_____ thousand _____ hundreds

Rename the number. Use the place-value chart to help.

5. 18 thousands = _____

THOUSANDS			ONES		
Hundreds	Tens	Ones	Hundreds	Tens	Ones

 6. 570,000 = 57 _____

THOUSANDS			ONES		
Hundreds	Tens	Ones	Hundreds	Tens	Ones

Rename the number.

7. 580 = _____ tens

8. 740,000 = _____ ten thousands

9. 8 hundreds 4 tens = 84 _____

10. 29 thousands = _____

🔑 UNLOCK the Problem ▸ REAL WORLD

11. A toy store is ordering 3,000 remote control cars. The store can order the cars in sets of 10. How many sets of 10 does the store need to order?

   Ⓐ 30   Ⓒ 3,000

   Ⓑ 300   Ⓓ 30,000

   a. What information do you need to use?

   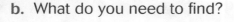

   _____

   b. What do you need to find?

   _____

   _____

   c. How can renaming numbers help you solve this problem?

   _____

   d. Describe a strategy you can use to solve the problem.

   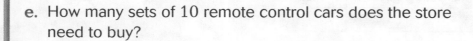

   _____

   e. How many sets of 10 remote control cars does the store need to buy?

   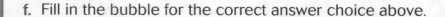

   f. Fill in the bubble for the correct answer choice above.

12. Adam sold 53 boxes of oranges during a citrus sale. There were 10 oranges in each box. How many oranges did he sell in all?

   Ⓐ 53   Ⓒ 5,300

   Ⓑ 530   Ⓓ 53,000

13. A store sold a total of 15,000 boxes of buttons last month. If the store sold 150,000 buttons, how many buttons were in each box?

   Ⓐ 10   Ⓒ 1,000

   Ⓑ 100   Ⓓ 10,000

Add Whole Numbers

Essential Question How can you add whole numbers?

 **UNLOCK the Problem** REAL WORLD

Alaska is the largest state in the United States by area. Its land area is 570,374 square miles and its water surface area is 86,051 square miles. Find the total area of Alaska.

- Underline what you are asked to find.
- Circle the information you will use.

 Find the sum.

Add. 570,374 + 86,051

Think: It is important to line up the addends by place value when adding two numbers.

STEP 1 Add the ones.

Add the tens. Regroup.

12 tens = 1 hundred _____ tens

$$\begin{array}{r} 5 7 0, \overset{1}{3} 7 4 \\ + \ 8 6, 0 5 1 \\ \hline \end{array}$$

▲ The area of Alaska is outlined in the photo above.

STEP 2 Add the hundreds.

Add the thousands.

$$\begin{array}{r} 5 7 0, \overset{1}{3} 7 4 \\ + \ 8 6, 0 5 1 \\ \hline 2 5 \end{array}$$

STEP 3 Add the ten thousands.

Regroup.

15 ten thousands =

1 hundred thousand _____ ten thousands

$$\begin{array}{r} \overset{1}{5} 7 0, \overset{1}{3} 7 4 \\ + \ 8 6, 0 5 1 \\ \hline 6, 4 2 5 \end{array}$$

Math Talk MATHEMATICAL PRACTICES
Explain how you know when to regroup when adding.

STEP 4 Add the hundred thousands.

$$\begin{array}{r} \overset{1}{5} 7 0, \overset{1}{3} 7 4 \\ + \ 8 6, 0 5 1 \\ \hline 5 6, 4 2 5 \end{array}$$

So, the total area of Alaska is _____ square miles.

Estimate You can estimate to tell whether an answer is reasonable.
To estimate a sum, round each addend before you add.

 Example **Estimate. Then find the sum.**

Juneau has an area of 2,717 square miles. Valdez has an area of
222 square miles. What is their combined area?

A Estimate. Use the grid to help you align the addends by place value.

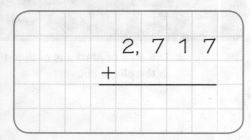

2, 7 1 7 →   3, 0 0 0   Round to the nearest thousand.

2 2 2 → +   Round to the nearest hundred.

So, the combined area of Juneau and Valdez is about _____
square miles.

B Find the sum.

     2, 7 1 7

  + _____

Think: Begin by adding the ones.

> **! ERROR Alert**
> Remember to align the
> addends by place value.

So, the combined area of Juneau and Valdez is _____
square miles.

• Is the sum reasonable? **Explain.**

_____

_____

Share and Show ....

1. Use the grid to find 738,901 + 162,389.

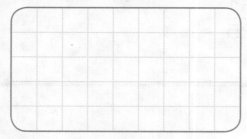

Use the grid to align the addends by place value.

Name _____

Problem Solving • Comparison Problems with Addition and Subtraction

Essential Question How can you use the strategy *draw a diagram* to solve comparison problems with addition and subtraction?

🔑 UNLOCK the Problem REAL WORLD

Hot air balloon festivals draw large crowds of people. The attendance on the first day of one festival was 17,350. On the second day the attendance was 18,925. How many more people attended the hot air balloon festival on the second day?

Use the graphic organizer to help you solve the problem.

Read the Problem

What do I need to find?	**What information do I need to use?**	**How will I use the information?**
Write what you need to find. _____ _____ _____	_____ people attended on the first day, _____ people attended on the second day.	What strategy can you use? _____ _____ _____

Solve the Problem

I can draw a bar model and write an equation to represent the problem.

18,925

17,350
⊔

18,925 − 17,350 = _____

So, _____ more people attended the festival on the second day.

🔑 Try Another Problem

During an event, a hot air balloon traveled a distance of 5,110 feet during the first trip and 850 feet more during the second trip. How far did it travel during the second trip?

Read the Problem

What do I need to find?	What information do I need to use?	How will I use the information?

Solve the Problem

- Is your answer reasonable? **Explain** how you know.

_____

_____

Math Talk
Explain how inverse operations can be used to check your answer.

Name _____

Share and Show

🕯 UNLOCK the Problem

✓ Use the Problem Solving MathBoard
✓ Underline important facts.
✓ Choose a strategy you know.

1. Hot air balloons are able to fly at very high altitudes. A world record height of 64,997 feet was set in 1988. In 2005, a new record of 68,986 feet was set. How many feet higher was the 2005 record than the 1988 record?

   First, draw a diagram to show the parts of the problem.

   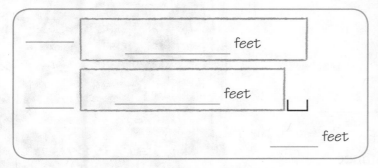

   _____ | feet
   _____ | feet
   _____ feet

▲ Dr. Vijaypat Singhania flew the world's largest hot-air balloon when he made his record-breaking flight. The balloon he flew was over 20 stories tall.

   Next, write the problem you need to solve.

   _____

   Last, solve the problem to find how many feet higher the 2005 record was than the 1988 record.

   So, the 2005 record was _____ feet higher.

2. **What if** a new world altitude record of 70,000 feet was set? How many feet higher would the new record be than the 2005 record?

   _____

✅ 3. Last year, the ticket sales for a commercial hot air balloon ride were $109,076. This year, the ticket sales were $125,805. How much more were the ticket sales this year?

   _____

✅ 4. There were 665 hot air balloon pilots at a hot air balloon race. There were 1,550 more ground crew members than there were pilots. How many ground crew members were there?

   _____

© Houghton Mifflin Harcourt Publishing Company

On Your Own . . . .

Choose a STRATEGY

Act It Out
Draw a Diagram
Find a Pattern
Make a Table or List
Solve a Simpler Problem

Use the information in the table for 5–7.

5. Steve Fossett attempted to fly around the world in a balloon several times before he succeeded in 2002. How many more miles did he fly during the 2002 flight than during the August 1998 flight?

PILOT

6. **H.O.T.** Is the combined distance for the 1998 flights more or less than the distance for the 2002 flight? **Explain.**

Steve Fossett's Balloon Flights	
Year	Distance in Miles
1996	2,200
1997	10,360
1998 (January)	5,803
1998 (August)	14,235
2001	3,187
2002	20,482

7. **Write Math** Estimate the total number of miles Fossett flew during the six hot air balloon flights. **Explain** how you estimated.

SHOW YOUR WORK

8. **Test Prep** Rusty wants to buy a small hot air balloon that costs $23,950. The cost of training for a license is $2,750. How much will Rusty pay for the balloon and the training?

Ⓐ $21,200   Ⓒ $26,700

Ⓑ $26,600   Ⓓ $36,700

FOR MORE PRACTICE: Standards Practice Book, pp. P17–P18

Name _____

 **Chapter Review/Test**

▶ **Vocabulary**

Choose the best term from the box.

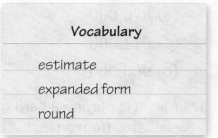
Vocabulary

estimate
expanded form
round

1. An _____ is close to an exact amount. (p. 17)

2. You can _____ to find an estimate. (p. 17)

▶ **Concepts and Skills**

Compare the values of the underlined digits.

3. 2,<u>4</u>02 and 6<u>4</u>,513

   The value of 4 in _____ is _____ times the value of 4 in _____.

Write the number in two other forms.

4. two hundred thirty-four thousand, one hundred sixty-four

   _____

   _____

5. 791,030

   _____

   _____

Compare. Write <, >, or =.

6. 600,849 _____ 398,989

7. 36,954 _____ 112,365

Round to the place of the underlined digit.

8. 62<u>4</u>,531

9. 4<u>6</u>3,356

10. <u>4</u>23,906

11. <u>5</u>83,342

   _____   _____   _____   _____

Rename the number.

12. 650 = _____ tens

13. 780,000 = 78 _____

Estimate. Then find the sum or difference.

14. Estimate: _____

$$\begin{array}{r} 185,239 \\ + 491,056 \\ \hline \end{array}$$

15. Estimate: _____

$$\begin{array}{r} 709,032 \\ - 249,136 \\ \hline \end{array}$$

Fill in the bubble completely to show your answer.

16. Pike National Forest located in California has a total area of 871,495 acres. What is the area to the nearest thousand?

Ⓐ 800,000

Ⓑ 870,000

Ⓒ 871,000

Ⓓ 900,000

17. Micah is playing a card game. To play, each person chooses six cards from a stack.

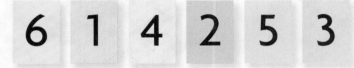

The player who makes the greatest six-digit number from the cards is the winner. What is the greatest number that can be made from the six cards shown?

Ⓐ 654,321

Ⓑ 365,124

Ⓒ 451,236

Ⓓ 563,412

18. Mr. Rodriguez bought 420 pencils for the school. If there are 10 pencils in a box, how many boxes of pencils did he buy?

Ⓐ 42   Ⓒ 4,200

Ⓑ 420   Ⓓ 42,000

19. Chan's website had 12,014 visitors and Pamela's website had 11,987 visitors. Kim's website had more visitors than Pamela's website, but fewer than Chan's website. Which of the following could be the number of visitors Kim's website had?

Ⓐ 13,001

Ⓑ 12,104

Ⓒ 12,001

Ⓓ 11,790

Name _____

Fill in the bubble completely to show your answer.

20. During the summer, the population of Spring Lake is 30,155. During the winter, the population drops down to 13,876. How many people spend only the summer months in Spring Lake?

Ⓐ 16,279

Ⓑ 24,207

Ⓒ 26,279

Ⓓ 44,031

21. The total attendance for the 2008 World Series of Baseball was 219,369. Which number below is greater than 219,369?

Ⓐ 209,369

Ⓑ 210,369

Ⓒ 218,369

Ⓓ 220,369

22. Which number rounded to the nearest hundred thousand is 800,000? Use the number line to help.

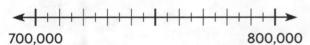

700,000   800,000

Ⓐ 164,328

Ⓑ 693,023

Ⓒ 750,012

Ⓓ 871,486

23. Theater attendance last year was 885,607 people. Which estimate is closest to the total number of people who attended performances last year?

Ⓐ 900,000

Ⓑ 800,000

Ⓒ 100,000

Ⓓ 90,000

▶ Constructed Response

24. Mt. Hunter has a height of 14,573 feet, Mt. McKinley has a height of 20,320 feet, and Mt. Whitney has a height of 14,505 feet. Name the mountains in order from least height to greatest height. Use pictures, words, or numbers to show how you know.

_____

_____

25. During September and October of 2008, the Grand Canyon National Park recorded a total of 792,426 visitors. If there were 359,396 visitors in October, how many people visited the park in September? Use pictures, words, or numbers to show how you know.

_____

_____

▶ Performance Task

26. Inez and Roy made three numbers with their number cards. Then their table got bumped and mixed up the cards. Look at the cards and help Inez and Roy make the three numbers again.

Ⓐ One number was the greatest six-digit number they could make. **Explain** how you found the greatest six-digit number.

_____

_____

_____

_____

_____

_____

Ⓑ Another number was the least five-digit number they could make.

_____

Ⓒ They had a four-digit number with a 5 in the thousands place and the ones place, a six in the tens place, and a 4 in the hundreds place.

_____

Multiply by 1-Digit Numbers

Show What You Know

Check your understanding of important skills.

Name _____

▶ **Arrays** Write a multiplication sentence for the array.

1.

2.

_____ _____   _____ _____

▶ **Multiplication Facts** Find the product.

3. _____ = 9 × 6   4. _____ = 7 × 8   5. 8 × 4 = _____

▶ **Regroup Through Thousands**
Regroup. Write the missing numbers.

6. 9 tens 10 ones = _____ hundred   7. 60 hundreds = _____ thousands

8. 25 tens = _____ hundreds 5 tens   9. 14 ones = _____ ten _____ ones

10. 3 tens 12 ones = _____ tens 2 ones

The Arctic Lion's Mane Jellyfish is one of the largest
known animals. Its tentacles can be as long as 120 feet.
Be a Math Detective to find how this length compares
to your height. Round your height to the nearest foot.
120 feet is _____ times as long as _____ feet.

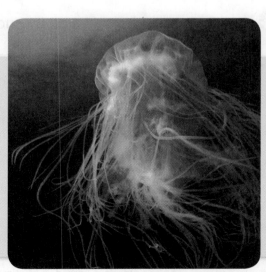

Vocabulary Builder

Review Words			Preview Words
✓ estimate	✓ place value	✓ rounding	Distributive Property
expanded form	product		partial product
factor	✓ regroup		

▶ Visualize It

Complete the flow map, using the words with a ✓.

Multiplying

What can you do?	What can you use?	What are some examples?
_____ products.	Use _____ and mental math.	3 × 48 = ■ ↓ ↓ 3 × 50 = 150
_____ ones as tens.	Use _____ .	12 ones = 1 ten 2 ones

▶ Understand Vocabulary

Complete the sentences.

1. The _____ states that multiplying a sum by a number is the same as multiplying each addend by the number and then adding the products.

2. A number that is multiplied by another number to find a product

   is called a _____ .

3. A method of multiplying in which the ones, tens, hundreds, and so on are multiplied separately and then the products are added

   together is called the _____ method.

GO Online • eStudent Edition • Multimedia eGlossary

Name _____

Multiplication Comparisons

Essential Question How can you model multiplication comparisons?

You can use multiplication to compare amounts.
For example, you can think of $15 = 3 \times 5$ as a comparison
in two ways:

Remember
The Commutative Property
states that you can multiply
two factors in any order and
get the same product.

15 is 3 times as many as 5.   *15 is 5 times as many as 3.*

15		
5	5	5

5

15				
3	3	3	3	3

3

🔑 UNLOCK the Problem REAL WORLD

Carly has 9 pennies. Jack has 4 times as many
pennies as Carly. How many pennies does
Jack have?

🔓 **Draw a model and write an equation
to solve.**

• **What do you need to compare?**

_____

_____

MODEL

Carly
____

Jack
____	____	____	____

RECORD

Use the model to write an equation and solve.

$n = $ _____ $\times$ _____

$n = $ _____

The value of n is 36.

Think: n is how many pennies Jack has.

So, Jack has _____ pennies.

MATHEMATICAL PRACTICES

Math Talk Describe what is
being compared and explain
how the comparison model
relates to the equation.

• **H.O.T.** **Explain** how the equation for *4 is 2 more than 2* is
different from the equation for *4 is 2 times as many as 2*.

_____

_____

_____

🔑 Example Draw a model and write an equation to solve.

Miguel has 3 times as many rabbits as Sara. Miguel has 6 rabbits. How many rabbits does Sara have?

- How many rabbits does Miguel have? _____
- How many rabbits does Sara have?

_____

_____

MODEL

Think: You don't know how many rabbits Sara has. Use *n* for Sara's rabbits.

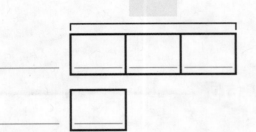

_____

_____

So, Sara has 2 rabbits.

RECORD

Use the model to write an equation and solve.

6 = _____ × _____

6 = 3 × _____   Think: 3 times what number equals 6?

The value of *n* is 2.

Think: *n* is how many rabbits Sara has.

Try This! Write an equation or a comparison sentence.

Ⓐ Write an equation.

21 is 7 times as many as 3.

_____ = _____ × _____

Ⓑ Write a comparison sentence.

8 × 5 = 40

_____ times as many as _____ is _____.

Share and Show 🖊️ MATH BOARD ....

1. There are 8 students in the art club. There are 3 times as many students in chorus. How many students are in chorus?

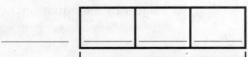

Write an equation and solve.

n = _____ × _____

n = _____

The value of *n* is _____.

So, there are _____ students in chorus.

MATHEMATICAL PRACTICES

Math Talk Could you write the equation a different way? Explain.

Name _____

Comparison Problems

Essential Question How does a model help you solve a comparison problem?

🔑 UNLOCK the Problem REAL WORLD

Evan's dog weighs 7 times as much as Oxana's dog. Together, the dogs weigh 72 pounds. How much does Evan's dog weigh?

🔑 Example 1 Use a multiplication model.

STEP 1 Draw a model. Let *n* represent the unknown.

> **Think:** Let *n* represent how much Oxana's dog weighs. Together, the dogs weigh 72 pounds.

Evan's

Oxana's

STEP 2 Use the model to write an equation. Find the value of *n*.

_____ × *n* = _____   **Think:** There are 8 parts. The parts together equal 72.

8 × _____ = 72   **Think:** What times 8 equals 72?

The value of *n* is 9.

n is how much _____ weighs.

STEP 3 Find how much Evan's dog weighs.

> **Think:** Evan's dog weighs 7 times as much as Oxana's dog.

Evan's dog = _____ × _____   Multiply.

= _____

So, Evan's dog weighs 63 pounds.

MATHEMATICAL PRACTICES
Math Talk Explain how you know you have found the weight of Evan's dog.

To find how many times as much, use a multiplication model. To find how many more or fewer, model the addition or subtraction.

Evan's dog weighs 63 pounds. Oxana's dog weighs 9 pounds. How much more does Evan's dog weigh than Oxana's dog?

🔒 Example 2 Use an addition or subtraction model.

STEP 1 Draw a model. Let n represent the unknown.

Think: Let n represent the difference.

_____ ┌──────────────────── _____ ────────────────────┐

_____ ┌─────┐
        │_____│ └──┘

_____

STEP 2 Use the model to write an equation. Find the value of n.

_____ − _____ = n **Think:** The model shows a difference.

63 − 9 = _____   Subtract.

The value of n is _____.

n is _____.

So, Evan's dog weighs 54 pounds more than Oxana's dog.

Share and Show 📝 MATH BOARD · · · · · · · · · · · · · · · · · ·

Math Talk Explain how you can choose a model to help solve a comparison problem.
MATHEMATICAL PRACTICES

1. Maria's dog weighs 6 times as much as her rabbit. Together the pets weigh 56 pounds. What does Maria's dog weigh?

Draw a model. Let n represent the unknown.

_____ ┌─────┬─────┬─────┬─────┬─────┬─────┐
        │_____│_____│_____│_____│_____│_____│ ┐
                                              │
_____ ┌─────┐   │
        │_____│ ──────────────────────────────┘

Write an equation to find the value of n. $7 \times n =$ _____. n is _____ pounds.

Multiply to find how much Maria's dog weighs. $8 \times 6 =$ _____

So, Maria's dog weighs _____ pounds.

Name _____

Draw a model. Write an equation and solve.

2. Last month Kim trained 3 times as many dogs as cats. If the total number of cats and dogs she trained last month is 28, how many cats did Kim train?

Draw a model.

Write an equation and solve.

_____

_____

_____

_____

3. How many more dogs than cats did Kim train?

Draw a model.

Write an equation and solve.

_____

_____

_____

_____

On Your Own . . . .

Practice: Copy and Solve Draw a model.

Write an equation and solve.

4. At the dog show, there are 4 times as many boxers as spaniels. If there are a total of 30 dogs, how many dogs are spaniels?

_____

5. There are 5 times as many yellow labs as terriers in the dog park. If there are a total of 18 dogs, how many dogs are terriers?

_____

6. Ben has 3 times as many guppies as goldfish. If he has a total of 20 fish, how many guppies does he have?

_____

7. Carlita saw 5 times as many robins as cardinals while bird watching. She saw a total of 24 birds. How many more robins did she see than cardinals?

_____

Problem Solving REAL WORLD

8. **Write Math** ▸ To get to a dog show, Mr. Luna first drives 7 miles west from his home and then 3 miles north. Next, he turns east and drives 11 miles. Finally, he turns north and drives 4 miles to the dog show. How far north of Mr. Luna's home is the dog show?

To solve the problem, Dara and Cliff drew diagrams. Which diagram is correct? **Explain**.

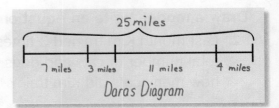

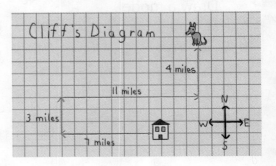

_____

_____

_____

_____

_____

9. Valerie and Bret have a total of 24 dog show ribbons. Bret has twice as many ribbons as Valerie. How many ribbons does each have?

_____

_____

10. **H.O.T.** Noah built a fenced dog run that is 8 yards long and 6 yards wide. He placed posts at every corner and every yard along the length and width of the run. How many posts did he use?

_____

11. **Test Prep** Brett and his mom paid $36 to attend a dog show. An adult ticket was 3 times the cost of a child's ticket. How much was an adult ticket?

Ⓐ $39

Ⓑ $36

Ⓒ $27

Ⓓ $9

SHOW YOUR WORK

© Houghton Mifflin Harcourt Publishing Company

Name _____

Multiply Tens, Hundreds, and Thousands

Essential Question How does understanding place value help you multiply tens, hundreds, and thousands?

UNLOCK the Problem REAL WORLD

Each car on a train has 200 seats. How many seats are on a train with 8 cars?

Find 8 × 200.

🔑 One Way Draw a quick picture.

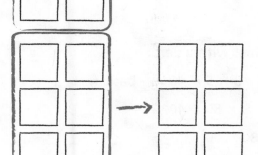

T

Think: 10 hundreds = 1,000

Think: 6 hundreds = 600

1,000 + 600 = _____

🔑 Another Way Use place value.

8 × 200 = 8 × _____ hundreds

= _____ hundreds

= _____   **Think:** 16 hundreds is 1 thousand, 6 hundreds.

So, there are _____ seats on a train with 8 cars.

Math Talk   MATHEMATICAL PRACTICES
Explain how finding 8 × 2 can help you find 8 × 200.

© Houghton Mifflin Harcourt Publishing Company

Other Ways

A Use a number line.

Bob's Sled Shop rents 4,000 sleds each month.
How many sleds does the store rent in 6 months?

Find 6 × 4,000.

Multiplication can be thought of as repeated addition.
Draw jumps to show the product.

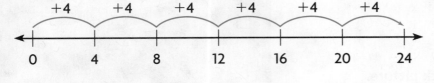

$6 \times 4 = 24$ ← basic fact

$6 \times 40 = 240$

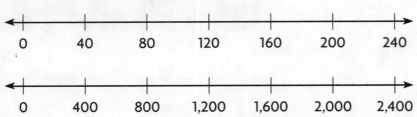

$6 \times 400 = 2,400$

$6 \times 4,000 = 24,000$

So, Bob's Sled Shop rents _____ sleds in 6 months.

B Use patterns.

Basic fact:

$3 \times 7 = 21$ ← basic fact

$3 \times 70 = 210$

$3 \times 700 =$ _____

$3 \times 7,000 =$ _____

Basic fact with a zero:

$8 \times 5 = 40$ ← basic fact

$8 \times 50 = 400$

$8 \times 500 =$ _____

$8 \times 5,000 =$ _____

- How does the number of zeros in the product of 8 and 5,000
  compare to the number of zeros in the factors? **Explain.**

_____

_____

Math Talk Describe how the number of zeros in
the factors and products changes in Example B.

Name _____

Share and Show

1. Use the drawing to find 2×500.

Math Talk MATHEMATICAL PRACTICES
Explain how to use place value to find 2×500.

$2 \times 500 =$ _____

Complete the pattern.

2. $3 \times 8 = 24$

$3 \times 80 =$ _____

$3 \times 800 =$ _____

$3 \times 8,000 =$ _____

3. $6 \times 2 = 12$

$6 \times 20 =$ _____

$6 \times 200 =$ _____

$6 \times 2,000 =$ _____

4. $4 \times 5 =$ _____

$4 \times 50 =$ _____

$4 \times 500 =$ _____

$4 \times 5,000 =$ _____

Find the product.

5. $6 \times 500 = 6 \times$ _____ hundreds

$=$ _____ hundreds

$=$ _____

6. $9 \times 5,000 = 9 \times$ _____ thousands

$=$ _____ thousands

$=$ _____

7. $9 \times 8,000 = 9 \times$ _____ thousands

$=$ _____ thousands

$=$ _____

8. $8 \times 7,000 = 8 \times$ _____ thousands

$=$ _____ thousands

$=$ _____

On Your Own

Find the product.

9. $7 \times 6,000 =$ _____

10. $4 \times 80 =$ _____

11. $6 \times 30 =$ _____

12. $9 \times 300 =$ _____

13. $5 \times 2,000 =$ _____

14. $3 \times 500 =$ _____

 Algebra Find the missing factor.

15. _____ $\times 9,000 = 63,000$

16. $7 \times$ _____ $= 56,000$

17. $8 \times$ _____ $= 3,200$

🔑 UNLOCK the Problem   REAL WORLD

18. Joe's Fun and Sun rents beach chairs. The store rented 300 beach chairs each month in April and in May. The store rented 600 beach chairs each month from June through September. How many beach chairs did the store rent during the 6 months?

- (A) 1,200
- (C) 3,000
- (B) 2,400
- (D) 5,400

a. What do you need to know? _____

_____

b. How will you find the number of beach chairs? _____

_____

c. Show the steps you use to solve the problem.

d. Complete the sentences.

For April and May, a total of _____ beach chairs were rented.

For June through September, a total of

_____ beach chairs were rented.

Joe's Fun and Sun rented _____ beach chairs during the 6 months.

e. Fill in the bubble for the correct answer choice above.

19. Carmen has three $20 bills and five $10 bills. How much money does she have?

- (A) $110
- (C) $60
- (B) $100
- (D) $50

20. Dan has 7 rolls of pennies. Each roll has 50 pennies. How many pennies does he have?

- (A) 57
- (C) 350
- (B) 300
- (D) 400

Name _____

Estimate Products

Essential Question How can you estimate products by rounding and determine if exact answers are reasonable?

 UNLOCK the Problem REAL WORLD

An elephant can reach as high as 23 feet with its trunk. It uses its trunk to pick up objects that weigh up to 3 times as much as a 165-pound person. About how much weight can an African elephant pick up with its trunk?

- Cross out the information you will not use.
- Circle the numbers you will use.
- How will you use the numbers to solve the problem?

_____

One Way Estimate by rounding.

STEP 1 Round the greater factor to the nearest hundred.

3×165

↓

3×200

STEP 2 Use mental math.

Think: $3 \times 200 = 3 \times 2$ hundreds

= 6 hundreds

= _____

So, an African elephant can pick up about 600 pounds with its trunk.

Another Way Estimate by finding two numbers the exact answer is between.

3×165 3×165

↓   ↓

$3 \times 100 =$ _____ $3 \times 200 =$ _____

Think: 165 is between 100 and 200. Use those numbers to estimate.

So, the African elephant can pick up between 300 and 600 pounds.

An African elephant is the largest living land mammal.

1. Is 200 less than or greater than 165? _____

2. So, would the product of 3 and 165 be less than or

greater than 600? _____

 Math Talk MATHEMATICAL PRACTICES
Is the exact answer closer to 300 or 600? Why?

Describe Reasonableness You can estimate a product to find whether an exact answer is reasonable.

 Tell whether an exact answer is reasonable.

Eva's horse eats 86 pounds each week. Eva solved the equation below to find how much feed she needs for 4 weeks.

$4 \times 86 = $

Eva says she needs 344 pounds of feed.
Is her answer reasonable?

One Way Estimate.

$$4 \times 86$$

↓ **Think:** Round to the nearest ten.

_____ × _____ = _____

344 is close to 360.

Another Way Find two numbers the exact answer is between.

$$4 \times 86$$

↓

_____ × _____ = _____

_____ is between _____ and _____.

$$4 \times 86$$

↓

_____ × _____ = _____

So, 344 pounds of feed is reasonable.

Share and Show

Math Talk MATHEMATICAL PRACTICES Is an exact answer of 11,065 reasonable? **Explain.**

1. Estimate the product by rounding.

$$5 \times 2,213$$

↓

_____ × _____ = _____

2. Estimate the product by finding two numbers the exact answer is between.

$$5 \times 2,213$$

_____ × _____ = _____

$$5 \times 2,213$$

_____ × _____ = _____

Name _____

Tell whether the exact answer is reasonable.

✓ 3. Kira needs to make color copies of a horse show flyer. The printer can make 24 copies in 1 minute. Kira says the printer makes 114 copies in 6 minutes.

_____

_____

_____

✓ 4. Jones Elementary is having a car wash to raise money for a community horse trail. Each car wash ticket costs $8. Tiara says the school will receive $1,000 if 125 tickets are sold.

_____

_____

_____

On Your Own

Tell whether the exact answer is reasonable.

5. Mrs. Hense sells a roll of coastal Bermuda horse hay for $58. She says she will make $174 if she sells 3 rolls.

_____

_____

_____

6. Mr. Brown sells horse supplies. A pair of riding gloves sells for $16. He says he will make $144 if he sells 9 pairs.

_____

_____

_____

7. A walking path for horses is 94 feet long. Carlos says that if a horse walks the length of the path 3 times, it will have walked 500 feet.

_____

_____

_____

8. **Test Prep** Which shows the two estimates that the exact answer is between?

$$4 \times 389$$

Ⓐ 300 and 400

Ⓑ 700 and 1,200

Ⓒ 600 and 1,000

Ⓓ 1,200 and 1,600

Make Predictions

As you read a story, you make predictions about what might happen next or about how the story will end.

When you solve a math problem, you make predictions about what your answer might be.

An *estimate* is a prediction because it helps you to determine whether your answer is correct. For some problems, it is helpful to make two estimates—one that is less than the exact answer and one that is greater.

H.O.T. **Predict whether the exact answer will be** *less than* **or** *greater than* **the estimate. Explain your answer.**

9. The food stand at the zoo sold 2,514 pounds of hamburger last month. The average cost of a pound of hamburger is $2. Jeremy estimates that about $6,000 worth of hamburger was sold last month.

_____

_____

_____

_____

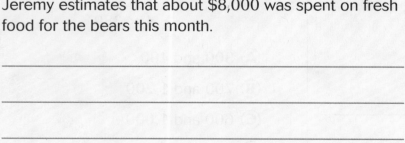

10. A zoo bought 2,240 pounds of fresh food for the bears this month. The average cost of a pound of food is $4. Jeremy estimates that about $8,000 was spent on fresh food for the bears this month.

_____

_____

_____

_____

Multiply Using the Distributive Property

Essential Question How can you use the Distributive Property to multiply a 2-digit number by a 1-digit number?

Investigate

Materials ■ color pencils, grid paper

You can use the Distributive Property to break apart numbers to make them easier to multiply.

The **Distributive Property** states that multiplying a sum by a number is the same as multiplying each addend by the number and then adding the products.

A. Outline a rectangle on the grid to model 6 × 13.

B. Think of 13 as 5 + 8. Break apart the model to show 6 × (5 + 8). Label and shade the smaller rectangles. Use two different colors.

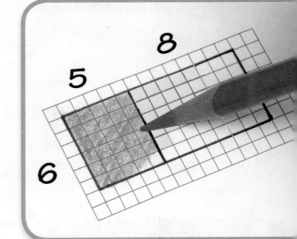

Use the Distributive Property. Find the product each smaller rectangle represents. Then find the sum of the products. Record your answers.

_____ × _____ = _____

_____ × _____ = _____

_____ + _____ = _____

C. Model 6 × 13 again. Think of 13 as a different sum. Break apart the model to show 6 × (_____ + _____). Find the product each smaller rectangle represents. Then find the sum of the products. Record your answers.

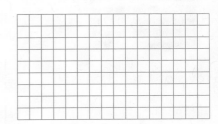

_____ × _____ = _____

_____ × _____ = _____

_____ + _____ = _____

Draw Conclusions

1. **Explain** how you found the total number of squares in each model in Steps B and C.

   _____

   _____

   _____

2. **Compare** the sums of the products in Steps B and C with those of your classmates. What can you conclude?

   _____

   _____

3. **H.O.T.** **Evaluate** To find 7 × 23, is it easier to break apart the factor, 23, as 20 + 3 or 15 + 8? Explain.

   _____

   _____

Make Connections

Another way to model the problem is to use base-ten blocks to show tens and ones.

STEP 1

Use base-ten blocks to model 6 × 13.

6 rows of 1 ten 3 ones

STEP 2

Break the model into tens and ones.

(6 × 1 ten)   (6 × 3 ones)

(6 × 10)   (6 × 3)

_____   _____

STEP 3

Add the tens and the ones to find the product.

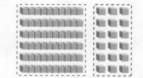

(6 × 10) + (6 × 3)

60   +   18

_____

So, 6 × 13 = 78.

In Step 2, the model is broken into two parts. Each part shows a **partial product**. The partial products are 60 and 18.

Math Talk   MATHEMATICAL PRACTICES

How does breaking apart the model into tens and ones make finding the product easier?

Name _____

Share and Show

Model the product on the grid. Record the product.

1. $3 \times 13 =$ _____

2. $5 \times 14 =$ _____

Find the product.

3. $6 \times 14 =$ _____

4. $5 \times 18 =$ _____

5. $4 \times 16 =$ _____

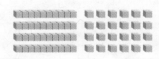

**Use grid paper or base-ten blocks to model the product.
Then record the product.**

6. $7 \times 12 =$ _____

7. $5 \times 16 =$ _____

8. $9 \times 13 =$ _____

9. $8 \times 11 =$ _____

10. $3 \times 15 =$ _____

11. $4 \times 12 =$ _____

12. $8 \times 18 =$ _____

13. $2 \times 19 =$ _____

14. $6 \times 17 =$ _____

15. $3 \times 19 =$ _____

16. $7 \times 15 =$ _____

17. $9 \times 16 =$ _____

18. **Write Math** **Explain** how modeling partial products can be
used to find the products of greater numbers.

_____

_____

_____

Problem Solving REAL WORLD

Pose a Problem

19. Kyle went to a fruit market. The market sells a wide variety of fruits and vegetables. The picture at the right shows a display of oranges.

Write a problem that can be solved using the picture.

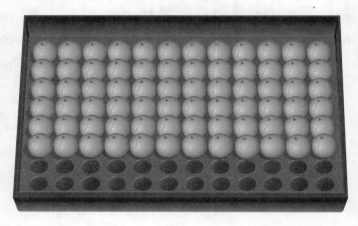

Pose a problem.

_____

_____

_____

_____

_____

_____

_____

Solve your problem.

_____

_____

_____

_____

_____

_____

_____

• Describe how you could change the problem by changing the number of rows of oranges and the number of empty spaces in the picture. Then solve the problem.

_____

_____

_____

_____

Multiply Using Expanded Form

Essential Question How can you use expanded form to multiply a multidigit number by a 1-digit number?

🔑 UNLOCK the Problem REAL WORLD

🔑 Example 1 Use expanded form.

Multiply. 5×143

$5 \times 143 = 5 \times ($ _____ $+$ _____ $+$ _____ $)$ Write 143 in expanded form.

$= (5 \times 100) + ($ _____ $\times$ _____ $) + ($ _____ $\times$ _____ $)$ Use the Distributive Property.

SHADE THE MODEL	THINK AND RECORD

STEP 1

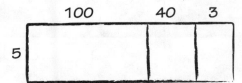

Multiply the hundreds.

$(5 \times 100) + (5 \times 40) + (5 \times 3)$

_____ $+ (5 \times 40) + (5 \times 3)$

STEP 2

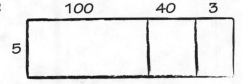

Multiply the tens.

$(5 \times 100) + (5 \times 40) + (5 \times 3)$

$500 \quad + $ _____ $+ (5 \times 3)$

STEP 3

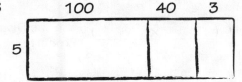

Multiply the ones.

$(5 \times 100) + (5 \times 40) + (5 \times 3)$

$500 \quad + \quad 200 \quad + $ _____

STEP 4

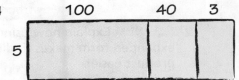

Add the partial products.

$$\begin{array}{r} 500 \\ 200 \\ +\ 15 \\ \hline \end{array}$$

So, $5 \times 143 =$ _____.

Math Talk MATHEMATICAL PRACTICES Is your answer reasonable? **Explain.**

Example 2 Use expanded form.

The gift shop at the animal park orders 3 boxes of toy animals. Each box has 1,250 toy animals. How many toy animals does the shop order?

Multiply. 3 × 1,250

STEP 1

Write 1,250 in expanded form. Use the Distributive Property.

3 × 1,250 = 3 × (_____ + _____ + _____)

= (3 × 1,000) + (_____ × _____) + (_____ × _____)

STEP 2

Add the partial products.

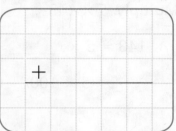

So, the shop ordered _____ animals.

Share and Show MATH BOARD · · · · · · · · · · · · · · · · · ·

1. Find 4 × 213. Use expanded form.

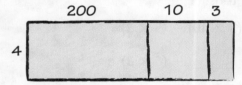

4 × 213 = _____ × (_____ + _____ + _____)

= (_____ × _____) + (_____ × _____) + (_____ × _____)   Use the Distributive Property.

= _____ + _____ + _____

= _____

Record the product. Use expanded form to help.

✅ **2.** 4 × 59 = _____

✅ **3.** 3 × 288 = _____

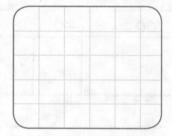

MATHEMATICAL PRACTICES

Math Talk **Explain** how using expanded form makes finding the product easier.

Name _____

On Your Own ....

Record the product. Use expanded form to help.

4. $4 \times 21 =$ _____

5. $6 \times 35 =$ _____

6. $5 \times 479 =$ _____

7. $8 \times 362 =$ _____

8. $7 \times 596 =$ _____

9. $2 \times 3,283 =$ _____

10. $4 \times 2,924 =$ _____

11. $6 \times 4,121 =$ _____

Problem Solving REAL WORLD

Use the table for 12–13.

Sacco Nursery Plant Sale		
Tree	Regular Price	Discounted Price (4 or more)
Flowering Cherry	$59	$51
Italian Cypress	$79	$67
Muskogee Crape Myrtle	$39	$34
Royal Empress	$29	$25

12. What is the total cost of 3 Italian cypress trees?

_____

13. **H.O.T.** **What's the Error?** Tanya says that the difference in the cost of 4 flowering cherry trees and 4 Muskogee crape myrtles is $80. Is she correct? **Explain.**

_____

_____

SHOW YOUR WORK

14. **Write Math** What is the greatest possible product of a 2-digit number and a 1-digit number? **Explain** how you know.

_____

_____

_____

15. **Test Prep** Which expression shows how to multiply 5×381 by using place value and expanded form?

(A) $(5 \times 3) + (5 \times 8) + (5 \times 1)$

(B) $(5 \times 300) + (5 \times 800) + (5 \times 100)$

(C) $(5 \times 300) + (5 \times 80) + (5 \times 1)$

(D) $(5 \times 300) + (5 \times 80) + (5 \times 10)$

FOR MORE PRACTICE:
Standards Practice Book, pp. P33–P34

Name _____

Multiply Using Partial Products

Essential Question How can you use place value and partial products to multiply by a 1-digit number?

 UNLOCK the Problem REAL WORLD

CONNECT How can you use what you know about the Distributive Property to break apart numbers to find products of 3-digit and 1-digit numbers?

🔑 **Use place value and partial products.**

Multiply. 6×182 **Estimate.** $6 \times 200 =$ _____

> • How can you write 182 as a sum of hundreds, tens, and ones?
>
> _____

	SHADE THE MODEL	THINK AND RECORD

STEP 1

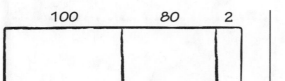

$$\begin{array}{r} 182 \\ \times\ 6 \\ \hline \end{array}$$

← Multiply the hundreds.
6×1 hundred $= 6$ hundreds

STEP 2

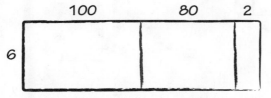

$$\begin{array}{r} 182 \\ \times\ 6 \\ \hline 600 \end{array}$$

← Multiply the tens.
6×8 tens $= 48$ tens

STEP 3

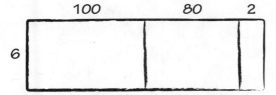

$$\begin{array}{r} 182 \\ \times\ 6 \\ \hline 600 \\ 480 \end{array}$$

← Multiply the ones.
6×2 ones $= 12$ ones

STEP 4

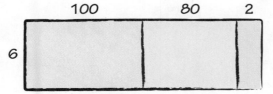

$$\begin{array}{r} 182 \\ \times\ 6 \\ \hline 600 \\ 480 \\ +\ 12 \\ \hline \end{array}$$

← Add the partial products.

So, $6 \times 182 = 1,092$. Since 1,092 is close to the estimate of 1,200, it is reasonable.

Math Talk MATHEMATICAL PRACTICES How can you use the Distributive Property to find 4×257?

🔒 Example

Use place value and partial products.

Multiply. 2 × 4,572.   **Estimate.** 2 × 5,000 = _____

$$\begin{array}{r} 4,572 \\ \times\ \ \ 2 \\ \hline \end{array}$$

← 2 × 4 thousands = 8 thousands
← 2 × 5 hundreds = 1 thousand
← 2 × 7 tens = 1 hundred, 4 tens
← 2 × 2 ones = 4 ones
← Add the partial products.

Share and Show ·

1. Use the model to find 2 × 137.

	100		30	7
2				

$$\begin{array}{r} 137 \\ \times\ \ \ 2 \\ \hline \\ \\ +\ \ \ \ \ \ \\ \hline \end{array}$$

Estimate. Then record the product.

2. Estimate: _____

$$\begin{array}{r} 190 \\ \times\ \ \ 3 \\ \hline \\ \\ +\ \ \ \ \ \ \\ \hline \end{array}$$

✅ **3.** Estimate: _____

$$\begin{array}{r} 471 \\ \times\ \ \ 4 \\ \hline \\ \\ +\ \ \ \ \ \ \\ \hline \end{array}$$

✅ **4.** Estimate: _____

$$\begin{array}{r} \$3,439 \\ \times\ \ \ 7 \\ \hline \\ \\ \\ +\ \ \ \ \ \ \\ \hline \end{array}$$

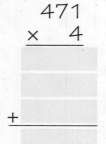

Math Talk MATHEMATICAL PRACTICES
Explain how using place value and expanded form makes it easier to find products.

Name _____

 Mid-Chapter Checkpoint

▶ **Vocabulary**

Choose the best term from the box to complete the sentence.

1. To find the product of a two-digit number and a 1-digit number, you can multiply the tens, multiply the ones, and find the sum of each _____. (p.62)

2. The _____ states that multiplying a sum by a number is the same as multiplying each addend by the number and then adding the products. (p.61)

▶ **Concepts and Skills**

Write a comparison sentence.

3. $5 \times 9 = 45$

_____ times as many as _____ is _____.

4. $24 = 6 \times 4$

_____ is _____ times as many as _____.

5. $54 = 6 \times 9$

_____ is _____ times as many as _____.

6. $8 \times 6 = 48$

_____ times as many as _____ is _____.

Estimate. Then record the product.

7. Estimate: _____

$$\begin{array}{r} 75 \\ \times\ 5 \\ \hline \end{array}$$

8. Estimate: _____

$$\begin{array}{r} 12 \\ \times\ 6 \\ \hline \end{array}$$

9. Estimate: _____

$$\begin{array}{r} 28 \\ \times\ 3 \\ \hline \end{array}$$

10. Estimate: _____

$$\begin{array}{r} \$43 \\ \times\ 6 \\ \hline \end{array}$$

Record the product. Use expanded form to help.

11. $5 \times 64 =$ _____

12. $3 \times 272 =$ _____

© Houghton Mifflin Harcourt Publishing Company

Fill in the bubble completely to show your answer.

13. There are 6 times as many dogs as cats. If the total number of dogs and cats is 21, how many dogs are there?

   Ⓐ 3

   Ⓑ 6

   Ⓒ 15

   Ⓓ 18

14. The table below shows the number of calories in 1 cup of different kinds of berries. How many calories are in 4 cups of blackberries?

Berry Nutrition	
Berry	Number of Calories in 1 Cup
Blackberries	62
Blueberries	83
Raspberries	64
Strawberries	46

   Ⓐ 62

   Ⓑ 83

   Ⓒ 248

   Ⓓ 308

15. The skating rink rents 200 pairs of skates in a month. How many pairs of skates does the rink rent in 4 months?

   Ⓐ 800

   Ⓑ 600

   Ⓒ 400

   Ⓓ 200

Multiply Using Mental Math

Essential Question How can you use mental math and properties to help you multiply numbers?

 UNLOCK the Problem REAL WORLD

Properties of Multiplication can make multiplication easier.

There are 4 sections of seats in the Playhouse Theater. Each section has 7 groups of seats. Each group has 25 seats. How many seats are there in the theater?

 Find 4 × 7 × 25.

4 × 7 × 25 = 4 × 25 × 7   Commutative Property

= _____ × 7   **Think:** 4 × 25 = 100

= _____   **Think:** 100 × 7 = 700

So, there are 700 seats in the theater.

25 seats ——

Stage

MATHEMATICAL PRACTICES

Math Talk How could knowing 4 × 25 help you find 6 × 25?

Try This! Use mental math and properties.

A **Find (6 × 10) × 10.**

(6 × 10) × 10 = 6 × (10 × 10)   Associative Property

= 6 × _____

= _____

B **Find (4 × 9) × 250.**

(4 × 9) × 250 = 250 × (4 × 9)   Commutative Property

= (250 × 4) × 9   Associative Property

= _____ × 9

= _____

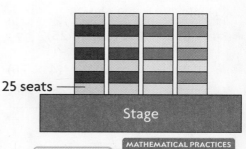

Remember

The Associative Property states that you can group factors in different ways and get the same product. Use parentheses to group the factors you multiply first.

More Strategies Choose the strategy that works best with the numbers in the problems.

🔑 Examples

A **Use friendly numbers.**

Multiply. 24×250

Think: $24 = 6 \times 4$ and $4 \times 250 = 1{,}000$

$24 \times 250 = 6 \times 4 \times 250$

$\qquad = 6 \times \underline{\hspace{1.5cm}}$

$\qquad = \underline{\hspace{1cm}}$

B **Use halving and doubling.**

Multiply. 16×50

Think: 16 can be divided evenly by 2.

$16 \div 2 = 8$ Find half of 16.

$8 \times 50 = \underline{\hspace{1.5cm}}$ Multiply.

$2 \times 400 = \underline{\hspace{1.5cm}}$ Double 400.

C **Use addition.**

Multiply. 4×625

Think: 625 is 600 plus 25.

$4 \times 625 = 4 \times (600 + 25)$

$\qquad = (4 \times 600) + (4 \times 25)$

$\qquad = \underline{\hspace{1.5cm}} + \underline{\hspace{1.5cm}}$

$\qquad = \underline{\hspace{1.5cm}}$

D **Use subtraction.**

Multiply. 5×398

Think: 398 is 2 less than 400.

$5 \times 398 = 5 \times (400 - 2)$

$\qquad = (5 \times \underline{\hspace{1.2cm}}) - (5 \times 2)$

$\qquad = 2{,}000 - \underline{\hspace{1.5cm}}$

$\qquad = \underline{\hspace{1.5cm}}$

- What property is being used in Examples C and D? _____

Share and Show

1. Break apart the factor 112 to find 7×112 by using mental math and addition.

   $7 \times 112 = 7 \times (\underline{\hspace{1.5cm}} + 12)$

   $= \underline{\hspace{5cm}}$

   $= \underline{\hspace{5cm}}$

   $= \underline{\hspace{5cm}}$

Name _____

Find the product. Tell which strategy you used.

2. $4 \times 6 \times 50$

_____

_____

⊘3. 5×420

_____

_____

⊘4. 6×298

_____

_____

On Your Own ·

Find the product. Tell which strategy you used.

Math Talk MATHEMATICAL PRACTICES
Explain how using an addition strategy is related to using a subtraction strategy.

5. 14×50

_____

_____

6. 32×25

_____

_____

7. $14 \times 25 \times 4$

_____

_____

8. $4 \times 15 \times 25$

_____

_____

9. 5×198

_____

_____

10. 5×250

_____

_____

Practice: Copy and Solve Use a strategy to find the product.

11. 16×400

12. $3 \times 31 \times 10$

13. 3×199

14. $3 \times 1,021$

 Algebra Use mental math to find the unknown number.

15. $21 \times 40 = 840$, so $21 \times 42 =$ _____.

16. $9 \times 60 = 540$, so $18 \times 30 =$ _____.

Problem Solving REAL WORLD

Use the table for 17–18.

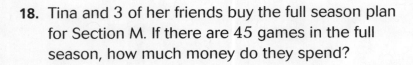

Arena Ticket Prices Per Game			
Section	Full Season	15-Game Plan	Gate Price
K	$44	$46	$48
L	$30	$32	$35
M	$25	$27	$30
N	$20	$22	$25

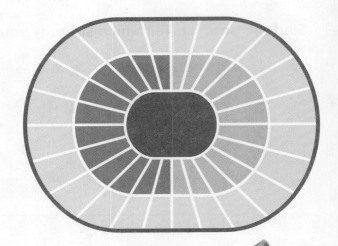

17. Three thousand, forty-three people buy tickets at the gate for Section N. How much money is collected for Section N at the gate?

_____

18. Tina and 3 of her friends buy the full season plan for Section M. If there are 45 games in the full season, how much money do they spend?

_____

19. **H.O.T.** **What's the Error?** Louisa says that 40 × 3,210 is 12,840. **Describe** and correct her error.

_____

_____

_____

_____

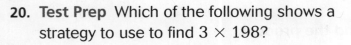
SHOW YOUR WORK

20. **Test Prep** Which of the following shows a strategy to use to find 3 × 198?

Ⓐ (3 × 200) − (3 × 2)

Ⓑ (3 × 200) + (3 × 2)

Ⓒ (3 × 198) − 6

Ⓓ 198 − 6

FOR MORE PRACTICE:
Standards Practice Book, pp. P37–P38

Name _____

Problem Solving • Multistep Multiplication Problems

Essential Question When can you use the *draw a diagram* strategy to solve a multistep multiplication problem?

UNLOCK the Problem REAL WORLD

At the sea park, one section in the stadium has 9 rows with 18 seats in each row. In the center of each of the first 6 rows, 8 seats are in the splash zone. How many seats are not in the splash zone?

Use the graphic organizer to help you solve the problem.

Read the Problem	Solve the Problem
What do I need to find? I need to find the number of seats that _____ in the splash zone.	I drew a diagram of the section to show 9 rows of 18 seats. In the center, I outlined a section to show the 6 rows of 8 seats in the splash zone.
What information do I need to use? There are 9 rows with _____ seats in each row of the section. There are 6 rows with _____ seats in each row of the splash zone.	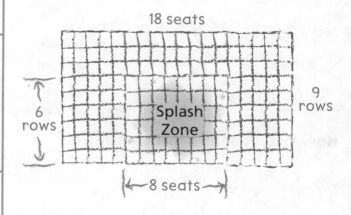
How will I use the information? I can _____ to find both the number of seats in the section and the number of seats in the splash zone.	

18
× 9
_____   ← total number of seats in the section

8
×6
_____   ← seats in the splash zone

1. What else do you need to do to solve the problem?

_____

_____

🔓 Try Another Problem

At the sea park, one section of the shark theater has 8 rows with 14 seats in each row. In the middle of the section, 4 rows of 6 seats are reserved. How many seats are not reserved?

Read the Problem	Solve the Problem
What do I need to find?	
What information do I need to use?	
How will I use the information?	

2. How did your diagram help you solve the problem?

_____

_____

MATHEMATICAL PRACTICES

Math Talk Explain how you can check your answer.

Name _____

Share and Show MATH BOARD

UNLOCK the Problem Tips

✓ Use the Problem Solving MathBoard.
✓ Underline important facts.
✓ Choose a strategy you know.

1. The seats in Sections A and B of the stadium are all taken for the last show. Section A has 8 rows of 14 seats each. Section B has 6 rows of 16 seats each. How many people are seated in Sections A and B for the last show?

   First, draw and label a diagram. **Next**, find the number of seats in each section.

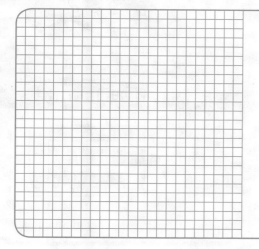

   Section A   Section B

   Last, find the total number of seats. _____ + _____ = _____

   There are _____ people seated in Sections A and B for the last show.

······ **SHOW YOUR WORK** ·······

2. **What if** Sections A and B each had 7 rows? How many people would have been seated in Sections A and B?

   _____

3. Carol, Ann, and Liz each bought a toy fish. Carol's fish is 10 inches longer than Ann's fish. Liz's fish is 2 inches longer than twice the length of Ann's fish. Ann's fish is 12 inches long. Find the length of each toy fish.

   _____

4. There are 8 rows of 22 chairs set up for an awards ceremony at the school. In each row, the 2 chairs on each end are reserved for students receiving awards. The rest of the chairs are for guests. How many chairs are there for guests?

   _____

On Your Own

Use the graph for 5–6.

5. Mr. Torres took his students to the dolphin show. Each row in the stadium had 11 seats. One adult sat at each end of a row, and each group of 4 students was seated between 2 adults. Mr. Torres sat by himself. How many adults were there?

Choose a STRATEGY

Act It Out
Draw a Diagram
Find a Pattern
Make a Table or List
Solve a Simpler Problem

6. **Write Math** Another stadium section has 24 rows of 10 seats each. Describe at least two ways Mrs. Allen's class can sit if an equal number of students sits in each row.

Sea Park Field Trips

7. Kari, Juan, Tami, and Brad are the first four people in line to see the Open Ocean exhibit. Kari is not first in line. Tami has at least two people ahead of her in line. Juan is third. Give the order of the first four people in line.

SHOW YOUR WORK

8. **H.O.T.** Nell made a secret code. Each code word has 2 letters. Each word begins with a consonant and ends with a vowel. How many code words can Nell make with 3 consonants and 2 vowels?

9. **Test Prep** A teacher has 29 students in her class. She gives each student 3 stickers and has no stickers left over. How many stickers did she have?

Ⓐ 67   Ⓑ 78   Ⓒ 87   Ⓓ 88

Name _____

Multiply 2-Digit Numbers with Regrouping

Essential Question How can you use regrouping to multiply a 2-digit number by a 1-digit number?

 UNLOCK the Problem REAL WORLD

A Thoroughbred racehorse can run at speeds of up to 60 feet per second. During practice, Celia's horse runs at a speed of 36 feet per second. How far does her horse run in 3 seconds?

- Underline important information.
- Is there information you will not use? If so, cross out the information.

Example 1

Multiply. 3×36 **Estimate.** $3 \times 40 =$ _____

MODEL	THINK	RECORD

STEP 1

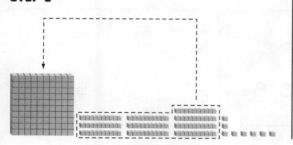

Multiply the ones.
3×6 ones = 18 ones
Regroup the 18 ones.

$$\begin{array}{r} 1 \\ 36 \\ \times\ 3 \\ \hline 8 \end{array}$$

Regroup 18 ones as 1 ten 8 ones.

STEP 2

Multiply the tens.
3×3 tens = 9 tens
Add the regrouped ten.
9 tens + 1 ten = 10 tens

$$\begin{array}{r} 1 \\ 36 \\ \times\ 3 \\ \hline 108 \end{array}$$

10 tens is the same as 1 hundred 0 tens.

So, Celia's racehorse runs _____ feet in 3 seconds.

Since _____ is close to the estimate of _____, the answer is reasonable.

Math Talk   MATHEMATICAL PRACTICES
Look at Step 1. **Explain** how the blocks show the regrouping of the 18 ones.

© Houghton Mifflin Harcourt Publishing Company

🔑 Example 2

Multiply. 8 × 22   **Estimate.** 8 × 20 = _____

MODEL	THINK	RECORD

STEP 1

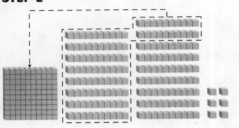

Multiply the ones.
8 × 2 ones = 16 ones

Regroup the 16 ones.

$$\begin{array}{r} 1 \\ 22 \\ \times\ 8 \\ \hline 6 \end{array}$$

Regroup
16 ones as
1 ten 6 ones.

STEP 2

Multiply the tens.
8 × 2 tens = 16 tens

Add the regrouped ten.
16 tens + 1 ten = 17 tens

$$\begin{array}{r} 1 \\ 22 \\ \times\ 8 \\ \hline 176 \end{array}$$

17 tens is
the same as 1
hundred 7 tens.

So, 8 × 22 = _____. Since _____ is close to the estimate

of _____, it is reasonable.

Try This! **Multiply.** 7 × $68

Estimate. 7 × $68	Use partial products.	Use regrouping.
	$ 6 8 × 7	$ 6 8 × 7

• Look at the partial products and regrouping methods above.

  How are the partial products 420 and 56 related to 476?

_____

Name _____

Share and Show

1. Use the model to find the product.

$2 \times 36 =$ _____

Estimate. Then record the product.

2. Estimate: _____

$$\begin{array}{r} 42 \\ \times\ 4 \\ \hline \end{array}$$

3. Estimate: _____

$$\begin{array}{r} 32 \\ \times\ 2 \\ \hline \end{array}$$

4. Estimate: _____

$$\begin{array}{r} 81 \\ \times\ 5 \\ \hline \end{array}$$

5. Estimate: _____

$$\begin{array}{r} \$63 \\ \times\ 7 \\ \hline \end{array}$$

Math Talk MATHEMATICAL PRACTICES
Describe the steps for using place value and regrouping to find 3×78.

On Your Own

Estimate. Then record the product.

6. Estimate: _____

$$\begin{array}{r} 33 \\ \times\ 2 \\ \hline \end{array}$$

7. Estimate: _____

$$\begin{array}{r} \$25 \\ \times\ 3 \\ \hline \end{array}$$

8. Estimate: _____

$$\begin{array}{r} 36 \\ \times\ 8 \\ \hline \end{array}$$

9. Estimate: _____

$$\begin{array}{r} \$94 \\ \times\ 5 \\ \hline \end{array}$$

Practice: Copy and Solve Estimate. Then record the product.

10. 3×82
11. 9×41
12. 6×75
13. $7 \times \$23$
14. $8 \times \$54$

15. 5×49
16. 8×97
17. 4×68
18. $9 \times \$68$
19. $6 \times \$73$

 Algebra Write a rule. Find the unknown numbers.

20.

Carton	_____	1	2	3	4	5
Eggs	_____	12	24		48	

21.

Row	_____	2	3	4	5	6
Seats	_____	32	48	64		

_____

_____

Problem Solving REAL WORLD

Use the table for 22–23.

22. At the speeds shown, how much farther could a black-tailed jackrabbit run than a desert cottontail in 7 seconds?

_____

23. A black-tailed jackrabbit hops about 7 feet in a single hop. How far can it hop in 5 seconds?

_____

Running Speeds	
Animal	**Speed (feet per second)**
Black-tailed Jackrabbit	51
Desert Cottontail	22

▲ Desert Cottontail

24. Mr. Wright bought a 3-pound bag of cat food and a 5-pound bag of dog food. There are 16 ounces in each pound. How many ounces of pet food did Mr. Wright buy?

_____

25. **H.O.T.** The sum of two numbers is 31. The product of the two numbers is 150. What are the numbers?

_____

26. **Write Math** 6×87 is greater than 5×87. How much greater? **Explain** how you know without multiplying.

_____

_____

_____

.... SHOW YOUR WORK ....

27. **Test Prep** Mrs. Sawyer bought a book for $25 and 3 toys for $13 each. How much change should she get back from a $100 bill?

(A) $26   (C) $46

(B) $36   (D) $56

FOR MORE PRACTICE:
Standards Practice Book, pp. P41–P42

Name _____

Multiply 3-Digit and 4-Digit Numbers with Regrouping

Essential Question How can you use regrouping to multiply?

🔑 UNLOCK the Problem REAL WORLD

Alley Spring, in Missouri, produces an average of 567 million gallons of water per week. How many gallons of water do the springs produce in 3 weeks?

Multiply. 3 × 567

Estimate. 3 × _____ = _____

THINK	RECORD

STEP 1

Multiply the ones.

3 × 7 ones = _____ ones
Regroup the 21 ones.

$$\begin{array}{r} \overset{2}{56}7 \\ \times\ \ 3 \\ \hline 1 \end{array}$$ Regroup the 21 ones as 2 tens and 1 one.

STEP 2

Multiply the tens.

3 × 6 tens = _____ tens
Add the regrouped tens.
18 tens + 2 tens = 20 tens
Regroup the 20 tens.

$$\begin{array}{r} \overset{2\,2}{56}7 \\ \times\ \ 3 \\ \hline 01 \end{array}$$ Regroup 20 tens as 2 hundreds 0 tens.

STEP 3

Multiply the hundreds.

3 × 5 hundreds = _____ hundreds
Add the regrouped hundreds.
15 hundreds + 2 hundreds = 17 hundreds

$$\begin{array}{r} \overset{2\,2}{56}7 \\ \times\ \ 3 \\ \hline 1,701 \end{array}$$ 17 hundreds is the same as 1 thousand 7 hundreds.

So, Alley Spring produces _____ gallons of water in 3 weeks.

🔑 Example

Use an estimate or an exact answer.

The table shows the prices of three vacation packages. Jake, his parents, and his sister want to choose a package.

Lakefront Vacations

	Adult	Child
Package A	$1,299	$619
Package B	$849	$699
Package C	$699	$484

Ⓐ About how much would Package C cost Jake's family?

STEP 1

Estimate the cost for 2 adults.

2 × $699

↓

2 × $700 = _____

STEP 2

Estimate the cost for 2 children.

2 × $484

↓

2 × $500 = _____

STEP 3

Add to estimate the total cost.

$$+ \underline{\hspace{2cm}}$$

So, Package C would cost Jake's family about $2,400.

Math Talk MATHEMATICAL PRACTICES
Explain how you know you can use an estimate.

Ⓑ Jake's family wants to compare the total costs of Packages A and C. Which plan costs more? How much more does it cost?

Package A

Adults	Children	Total Cost
$1,299	$619	
× 2	× 2	+

Package C

Adults	Children	Total Cost
$699	$484	
× 2	× 2	+

Subtract to compare the total costs of the packages.

$$\$3,836$$
$$- \$2,366$$
$$\overline{\hspace{3cm}}$$

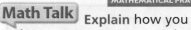
Math Talk MATHEMATICAL PRACTICES
Explain why you need an exact answer.

So, Package _____ would cost _____ more than Package _____.

Name _____

Share and Show

1. Tell what is happening in Step 1 of the problem.

_____

_____

STEP 1	STEP 2	STEP 3	STEP 4
2	4 2	1 4 2	1 4 2
1,274	1,274	1,274	1,274
× 6	× 6	× 6	× 6
4	44	644	7,644

Estimate. Then find the product.

2. Estimate: _____

$$603 \times 4$$

☑ 3. Estimate: _____

$$1,935 \times 7$$

☑ 4. Estimate: _____

$$\$8,326 \times 5$$

Math Talk MATHEMATICAL PRACTICES
Explain how you can use estimation to find how many digits the product 4 × 1,861 will have.

On Your Own

Estimate. Then find the product.

5. Estimate: _____

$$\$3,316 \times 8$$

6. Estimate: _____

$$\$2,900 \times 7$$

7. Estimate: _____

$$\$4,123 \times 6$$

8. Estimate: _____

$$\$1,893 \times 4$$

9. Estimate: _____

$$\$9,042 \times 8$$

10. Estimate: _____

$$3,286 \times 5$$

H.O.T. **Practice: Copy and Solve** Compare. Write <, >, or =.

11. 5 × 352 ◯ 4 × 440

12. 6 × 8,167 ◯ 9,834 × 5

13. 3,956 × 4 ◯ 5 × 7,692

14. 740 × 7 ◯ 8 × 658

15. 4 × 3,645 ◯ 5 × 2,834

16. 6,573 × 2 ◯ 4,365 × 3

Problem Solving  REAL WORLD

17. Look at the table. About how many more people visited the park in 2007 than in 2000?

Table Rock State Park Attendance	
Year	Number of Visitors
2000	869,736
2007	1,160,031

_____

18. Philadelphia, Pennsylvania, is 2,147 miles from Salt Lake City, Utah, and 2,868 miles from Portland, Oregon. What is the difference in the round-trip distances between Philadelphia and each of the other two cities? **Explain** whether you need an estimate or an exact answer.

_____

_____

.... SHOW YOUR WORK ....

19. **H.O.T.** **Sense or Nonsense?** Joe says that the product of a 4-digit number and a 1-digit number is always a 4-digit number. Does Joe's statement make sense? **Explain**.

_____

_____

_____

_____

20. **Test Prep** What number is 150 more than the product of 5 and 4,892?

- (A) 24,610
- (B) 24,160
- (C) 24,061
- (D) 25,610

Name _____

Solve Multistep Problems Using Equations

Essential Question How can you represent and solve multistep problems using equations?

 **UNLOCK the Problem** REAL WORLD

Crismari's computer has 3 memory cards with 64 gigabytes of space each and 2 memory cards with 16 gigabytes of space each. The files on her computer use 78 gigabytes of space. How much memory does her computer have left?

• Underline the important information.

 One Way Use multiple single-step equations.

STEP 1 Find how much memory is on 3 memory cards with 64 gigabytes of space each.

| 64 | 64 | 64 |

← 3 cards with 64 gigabytes.

n ← Total memory on 3 cards with 64 gigabytes.

$3 \times 64 = n$

_____ $= n$

STEP 2 Find how much memory is on 2 memory cards with 16 gigabytes of space.

| 16 | 16 |

← 2 cards with 16 gigabytes.

p ← Total memory on 2 cards with 16 gigabytes.

$2 \times 16 = p$

_____ $= p$

STEP 3 Find the total memory on the computer.

Total memory on 64-gigabyte cards.   Total memory on 16-gigabyte cards.

| 192 | 32 |

A ← Total memory on computer.

$192 + 32 = A$

_____ $= A$

STEP 4 The files use 78 gigabytes of space. Find how much memory the computer has left.

memory left   memory used

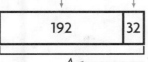

| y | 78 |

224 ← Total memory on the computer.

$224 - 78 = y$

_____ $= y$

So, Crismari has _____ gigabytes of memory left on her computer.

Order of Operations The Order of Operations is a special set of rules that gives the order in which calculations are done in an expression. First, multiply and divide from left to right. Then, add and subtract from left to right.

 Another Way Use one multistep equation.

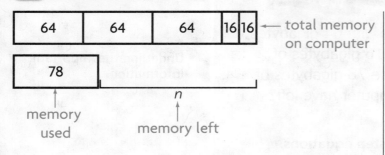

$3 \times 64 + 2 \times 16 - 78 = n$

_____ + _____ × _____ − _____ = n

_____ + _____ − _____ = n

_____ − _____ = n

_____ = n

Share and Show MATH BOARD . . . .

1. Use the order of operations to find the value of n.

   $5 \times 17 + 5 \times 20 - 32 = n$

   _____ + _____ × _____ − _____ = n ← First, multiply 5×17.

   _____ + _____ − _____ = n ← Next, multiply 5×20.

   _____ − _____ = n ← Then, add the two products.

   _____ = n ← Finally, subtract to find n.

Find the value of n.

2. $3 \times 22 + 7 \times 41 - 24 = n$

   _____ = n

3. $4 \times 34 + 6 \times 40 - 66 = n$

   _____ = n

4. $2 \times 62 + 8 \times 22 - 53 = n$

   _____ = n

5. $6 \times 13 + 9 \times 34 - 22 = n$

   _____ = n

Math Talk MATHEMATICAL PRACTICES
If you add before multiplying, will you get the same answer? **Explain**.

On Your Own

Find the value of *n*.

6. $8 \times 42 + 3 \times 59 - 62 = n$

_____ $= n$

7. $6 \times 27 + 2 \times 47 - 83 = n$

_____ $= n$

Problem Solving REAL WORLD

8. Maggie has 3 binders with 25 stamps in each binder. She has 5 binders with 24 baseball cards in each binder. If she gives 35 stamps to a friend, how many stamps and cards does she have left?

_____

9. Maddox has 4 boxes with 32 marbles in each box. He has 7 boxes with 18 shells in each box. If he gets 20 marbles from a friend, how many marbles and shells does he have?

_____

10. Test Prep Trina has 2 bags with 14 pinecones in each bag. She has 7 boxes with 15 acorns in each box. If she trades 5 pinecones for 10 acorns, how many pinecones and acorns does she have?

Ⓐ 28 Ⓒ 133

Ⓑ 105 Ⓓ 138

H.O.T. **What's the Error?**

11. Dominic has 5 books with 12 postcards in each book. He has 4 boxes with 20 coins in each box. If he gives 15 post cards to a friend, how many postcards and coins does he have?

Dominic drew this model.

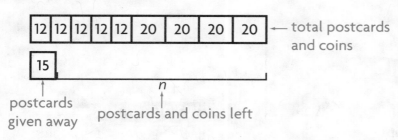

| 12 | 12 | 12 | 12 | 12 | 20 | 20 | 20 | 20 | ← total postcards and coins

| 15 |

postcards given away

n

postcards and coins left

Dominic used these steps to solve.

$5 \times 12 + 4 \times 20 - 15 = n$

$60 + 4 \times 20 - 15 = n$

$64 \times 20 - 15 = n$

$1,280 - 15 = n$

$1,265 = n$

Look at the steps Dominic used to solve this problem. Find and describe his error.

Use the correct steps to solve the problem.

So, there are _____ postcards and coins left.

Name _____

 **Chapter Review/Test**

▶ **Vocabulary**

Choose the best term from the box.

Vocabulary
Commutative Property
Distributive Property
partial product

1. To find the product of a 3-digit number and a 1-digit number, you can multiply the ones, multiply the tens, multiply the

   hundreds, and find the sum of each _____.
   (p. 62)

2. The _____ states that multiplying a sum by a number is the same as multiplying each addend by the number and then adding the products. (p. 61)

▶ **Concepts and Skills**

Estimate. Then find the product.

3. Estimate: _____

$$\begin{array}{r} 55 \\ \times\ \ 2 \\ \hline \end{array}$$

4. Estimate: _____

$$\begin{array}{r} \$25 \\ \times\ \ 3 \\ \hline \end{array}$$

5. Estimate: _____

$$\begin{array}{r} 306 \\ \times\ \ 8 \\ \hline \end{array}$$

6. Estimate: _____

$$\begin{array}{r} \$924 \\ \times\ \ \ \ 5 \\ \hline \end{array}$$

7. Estimate: _____

$$\begin{array}{r} 3,563 \\ \times\ \ \ \ 9 \\ \hline \end{array}$$

8. Estimate: _____

$$\begin{array}{r} 7,048 \\ \times\ \ \ \ 7 \\ \hline \end{array}$$

9. Estimate: _____

$$\begin{array}{r} 6,203 \\ \times\ \ \ \ 3 \\ \hline \end{array}$$

10. Estimate: _____

$$\begin{array}{r} 8,798 \\ \times\ \ \ \ 6 \\ \hline \end{array}$$

Fill in the bubble completely to show your answer.

11. Which number sentence shows the Distributive Property?

Ⓐ $2 \times 3 = 3 \times 2$

Ⓑ $5 \times 0 = 0$

Ⓒ $3 \times (5 + 2) = (3 \times 5) + (3 \times 2)$

Ⓓ $(3 \times 7) \times 4 = 3 \times (7 \times 4)$

12. Look at the pattern below. What is the missing number?

$$5 \times 6 = 30$$
$$5 \times 60 = 300$$
$$5 \times 600 = 3,000$$
$$5 \times \blacksquare = 30,000$$

Ⓐ 8,000

Ⓑ 6,000

Ⓒ 600

Ⓓ 60

13. Which comparison sentence represents the equation?

$$45 = 5 \times 9$$

Ⓐ 9 more than 5 is 45.

Ⓑ 9 is 5 times as many as 45.

Ⓒ 5 is 4 times as many as 45.

Ⓓ 45 is 5 times as many as 9.

14. There are 4 times as many alligators as crocodiles. If the total number of alligators and crocodiles is 40, how many alligators are there?

Ⓐ 40

Ⓑ 32

Ⓒ 24

Ⓓ 8

Name _____

Fill in the bubble completely to show your answer.

15. Gardeners at Seed Stop are planting seeds in 12-row seed trays. They plant 8 seeds in each row. How many plants will there be in each tray if all of the seeds germinate, or grow?

Ⓐ 84　　　　　　Ⓒ 96

Ⓑ 86　　　　　　Ⓓ 104

16. Which shows the product of $4 \times 15 \times 25$?

Ⓐ 150　　　　　　Ⓒ 1,500

Ⓑ 1,200　　　　　Ⓓ 1,600

17. A Broadway musical group will have 9 performances. The theater can seat 2,518 people. If all of the seats at each performance are taken, how many people will see the show?

Ⓐ 18,592　　Ⓒ 22,662

Ⓑ 22,652　　Ⓓ 31,622

18. The table below shows the type of film sold and the number of rolls in one pack at a local gift shop.

Gift Shop Film
Type of Film (pack of 4 rolls)
36 exposures
24 exposures
12 exposures

Hannah buys 3 packs of 36 exposure film and 2 packs of 24 exposure film. She uses 8 rolls of film. How many rolls does she have left?

Ⓐ 8　　　　　　Ⓒ 20

Ⓑ 12　　　　　　Ⓓ 24

19. John's grade has 3 classrooms. Each classroom has 14 tables. Two students sit at each table. About how many students are there in all?
Use pictures, words, or numbers to show how you know.

_____

▶ **Performance Task**

20. Justin has $450 to buy supplies for the school computer lab. He buys 8 boxes of printer paper that cost $49 each.

Ⓐ About how much money does Justin spend on the printer paper? **Describe** how you made your estimate.

_____

_____

_____

Ⓑ Find the actual amount of money Justin spends on the printer paper. **Explain** whether your estimate is close to the actual price.

_____

_____

_____

Ⓒ Will Justin have enough money left over to buy 3 packages of blank DVDs that cost $17 each? **Explain** your answer.

_____

_____

_____

3 Multiply 2-Digit Numbers

Show What You Know ✓

Check your understanding of important skills.

Name _____

▶ **Practice Multiplication Facts** Find the product.

1. 8 × 7 = _____

7 × 8 = _____

2. 3 × (2 × 4) = _____

(3 × 2) × 4 = _____

▶ **2-Digit by 1-Digit Multiplication** Find the product.

3. 28
× 3

4. 56
× 6

5. 71
× 5

6. 69
× 8

7. 36
× 4

▶ **Multiply by 1-Digit Numbers** Find the product.

8. 72
× 4

9. 456
× 5

10. 804
× 7

11. 1,341
× 9

12. 65
× 6

13. 392
× 8

14. 1,478
× 3

15. $1,627
× 2

16. 584
× 7

17. 2,837
× 4

MATH DETECTIVE WITH CARMEN SANDIEGO™

Yellowstone National Park, which is located in Wyoming, Montana, and Idaho, was America's first National Park. The park has over 500 geysers. Grand Geyser erupts about every 8 hours.

Be a Math Detective. Based on this estimate, how many times would you see this geyser erupt if you could watch it for 1 year? There are 24 hours in a day and 365 days in a year.

Vocabulary Builder

▶ **Visualize It** •••••••••••••••••••••••••••••

Complete the H-diagram using the words with a ✓.

Multiplication Words	Estimation Words

Review Words

Associative Property of
Multiplication
Commutative Property
of Multiplication
✓ estimate
✓ factor
✓ partial product
✓ place value
✓ product
regroup
✓ round

Preview Words

✓ compatible numbers

▶ **Understand Vocabulary** •••••••••••••••••••••••••••••

Draw a line to match each word or phrase with its definition.

Word

Definition

1. Commutative Property of Multiplication

2. estimate

3. compatible numbers

4. factor

5. regroup

- A number that is multiplied by another number to find a product

- To exchange amounts of equal value to rename a number

- To find an answer that is close to the exact amount

- Numbers that are easy to compute mentally

- The property that states when the order of two factors is changed, the product is the same.

© Houghton Mifflin Harcourt Publishing Company

Multiply by Tens

Essential Question What strategies can you use to multiply by tens?

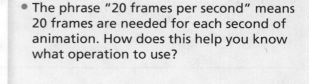

🔑 UNLOCK the Problem REAL WORLD

Animation for a computer-drawn cartoon requires about 20 frames per second. How many frames would need to be drawn for a 30-second cartoon?

- The phrase "20 frames per second" means 20 frames are needed for each second of animation. How does this help you know what operation to use?

_____

_____

One Way Use place value.

Multiply. 20×30

You can think of 30 as 3 tens.

$20 \times 30 = 20 \times$ _____ tens

$\qquad = $ _____ tens

$\qquad = 600$

Another Way Use the Associative Property.

You can think of 30 as 3×10.

$20 \times 30 = 20 \times (3 \times 10)$

$\qquad = (20 \times 3) \times 10$

$\qquad = $ _____ $\times$ _____

$\qquad = $ _____

So, _____ frames would need to be drawn.

Remember

The Associative Property states that you can group factors in different ways and get the same product. Use parentheses to group the factors you multiply first.

MATHEMATICAL PRACTICES

Math Talk How can you use place value to tell why $60 \times 10 = 600$? **Explain.**

- Compare the number of zeros in each factor to the number of zeros in the product. What do you notice?

_____

_____

🔑 Other Ways

A Use a number line and a pattern to multiply 15 × 20.

Draw jumps to show the product.

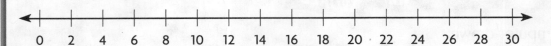

15 × 2 = _____

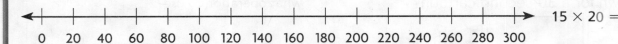

15 × 20 = _____

B Use mental math to find 14 × 30.

Use the halving-and-doubling strategy.

STEP 1 Find half of 14 to make the problem simpler.	**STEP 2** Multiply.	**STEP 3** Double 210.
Think: To find half of a number, divide by 2.		**Think:** To double a number, multiply by 2.
14 ÷ 2 = _____	7 × 30 = _____	2 × 210 = _____

So, 14 × 30 = 420.

Try This! Multiply.

Use mental math to find 12 × 40.

Use place value to find 12 × 40.

Share and Show ·

1. Find 20 × 27. Tell which method you chose. **Explain** what happens in each step.

_____

_____

Name _____

Choose a method. Then find the product.

2. 10×12

3. 20×20

✓**4.** 40×24

✓**5.** 11×60

> **Math Talk** MATHEMATICAL PRACTICES
> Explain how you can use
> $30 \times 10 = 300$ to find 30×12.

On Your Own ....

Choose a method. Then find the product.

6. 70×55

7. 17×30

8. 49×50

9. 10×70

10. 20×29

11. 50×46

12. 30×60

13. 12×90

 Algebra Find the unknown digit in the number.

14. $64 \times 40 = 2{,}56$ ■

15. $29 \times 50 = 1{,}$⬠50

16. 3◆$\times 47 = 1{,}410$

■ = _____

⬠ = _____

◆ = _____

Problem Solving  REAL WORLD

Use the table for 17–18.

17. How many frames did it take to produce 50 seconds of *Pinocchio*?

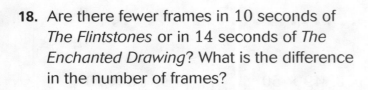

Animated Productions

Title	Date Released	Frames per Second
The Enchanted Drawing©	1900	20
Little Nemo©	1911	16
Snow White and the Seven Dwarfs©	1937	24
Pinocchio©	1940	19
The Flintstones™	1960–1966	24

_____

18. Are there fewer frames in 10 seconds of *The Flintstones* or in 14 seconds of *The Enchanted Drawing*? What is the difference in the number of frames?

_____

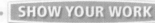 **SHOW YOUR WORK**

19. **H.O.T.** The product of my number and twice my number is 128. What is half my number? **Explain** how you solved the problem.

_____

_____

_____

20. **H.O.T. What's the Error?** Tanya says that the product of a multiple of ten and a multiple of ten will always have only one zero. Is she correct? **Explain**.

_____

_____

21. **Test Prep** Luis jogs 10 miles a week. He bikes 20 miles a week. How far will he have jogged in 26 weeks?

Ⓐ 30 miles   Ⓒ 260 miles

Ⓑ 200 miles   Ⓓ 520 miles

Name _____

Estimate Products

Essential Question What strategies can you use to estimate products?

UNLOCK the Problem REAL WORLD

The Smith family opens the door of their refrigerator 32 times in one day. There are 31 days in May. About how many times is it opened in May?

• Underline any information you will need.

One Way Use rounding and mental math.

Estimate. 32×31

STEP 1 Round each factor.

32×31

$\downarrow \quad \downarrow$

30×30

STEP 2 Use mental math.

$3 \times 3 = 9 \leftarrow$ basic fact

$30 \times 30 = $ _____

So, the Smith family opens the refrigerator door about 900 times during the month of May.

MATHEMATICAL PRACTICES

Math Talk Will the actual number of times the refrigerator is opened in a year be greater than or less than 900? **Explain.**

1. On average, a refrigerator door is opened 38 times each day. About how many fewer times in May is the Smith family's refrigerator door opened than the average refrigerator door?

 Show your work.

All 24 light bulbs in the Park family's home are CFL light bulbs. Each CFL light bulb uses 28 watts to produce light. About how many watts will the light bulbs use when turned on all at the same time?

🔑 Another Way Use mental math and compatible numbers.

Compatible numbers are numbers that are easy to compute mentally.

Estimate. 24 × 28

STEP 1 Use compatible numbers.

24 × 28
↓   ↓
25 × 30   Think: 25 × 3 = 75

So, about 750 watts are used.

STEP 2 Use mental math.

25 × 3 = 75

25 × 30 = _____

Try This! Estimate 26 × $79.

Ⓐ Round to the nearest ten	Ⓑ Compatible numbers
26 × $79	26 × $79 Think: How can you use 25 × 4 = 100 to help find 25 × 8?
↓ ↓	↓ ↓
_____ × _____ = _____	25 × $80 = _____
26 × $79 is about _____.	26 × $79 is about _____.

2. Explain why $2,400 and $2,000 are both reasonable estimates.

_____

_____

3. In what situation might you choose to find an estimate rather than an exact answer?

_____

_____

Share and Show . . . .

1. To estimate the product of 62 and 28 by rounding, how would you round the factors? What would the estimated product be?

_____

Name _____

Estimate the product. Choose a method.

2. 96×34

3. $47 \times \$39$

4. 78×72

Math Talk Describe how you know if an estimated product will be greater than or less than the exact answer.

MATHEMATICAL PRACTICES

On Your Own ....

Estimate the product. Choose a method.

5. 41×78

6. 51×73

7. 34×80

8. 84×23

9. $27 \times \$56$

10. 45×22

Practice: Copy and Solve Estimate the product. Choose a method.

11. 61×31

12. 52×68

13. 26×44

14. $57 \times \$69$

15. 55×39

16. 51×81

17. $47 \times \$32$

18. 49×64

 Find two possible factors for the estimated product.

19. 2,800

20. 8,100

21. 5,600

22. 2,400

_____

_____

© Houghton Mifflin Harcourt Publishing Company

Problem Solving REAL WORLD

23. On average, a refrigerator door is opened 38 times each day. Len has two refrigerators in his house. Based on this average, about how many times in a 3-week period are the refrigerator doors opened?

_____

24. The cost to run a refrigerator is about $57 each year. About how much will it have cost to run by the time it is 15 years old?

_____

25. If Mel opens his refrigerator door 36 times every day, about how many times will it be opened in April? Will the exact answer be more than or less than the estimate? **Explain**.

_____

_____

_____

_____

_____

26. **H.O.T.** **What's the Question?** The estimated product of two numbers, that are not multiples of ten, is 2,800.

_____

_____

27. **Test Prep** Which is the best estimate for the product 75×23?

Ⓐ 2,600   Ⓒ 1,600

Ⓑ 2,200   Ⓓ 160

Name _____

Area Models and Partial Products

Essential Question How can you use area models and partial products to multiply 2-digit numbers?

Investigate

Materials ■ color pencils

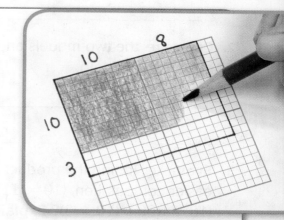

How can you use a model to break apart factors and make them easier to multiply?

A. Outline a rectangle on the grid to model 13×18. Break apart the model into smaller rectangles to show factors broken into tens and ones. Label and shade the smaller rectangles. Use the colors below.

B. Find the product of each smaller rectangle. Then, find the sum of the partial products. Record your answers.

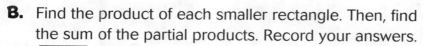 = 10×10

= 10×8

= 3×10

= 3×8

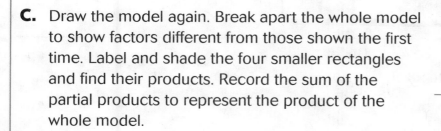

100 + ____ + ____ + ____ = ____

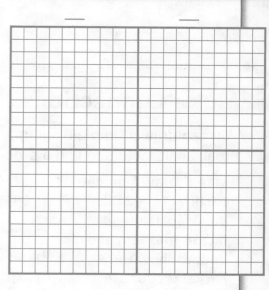

C. Draw the model again. Break apart the whole model to show factors different from those shown the first time. Label and shade the four smaller rectangles and find their products. Record the sum of the partial products to represent the product of the whole model.

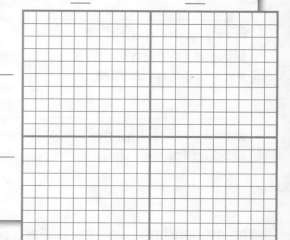

_____ + _____ + _____ + _____ = _____

Draw Conclusions

1. **Explain** how you found the total number of squares in the whole model.

   _____

   _____

2. **Compare** the two models and their products. What can you conclude? **Explain.**

   _____

   _____

3. **Evaluate** To find the product of 10 and 33, which is the easier computation, $(10 \times 11) + (10 \times 11) + (10 \times 11)$ or $(10 \times 30) + (10 \times 3)$? **Explain.**

   _____

   _____

Make Connections

You can draw a simple diagram to model and break apart factors to find a product. Find 15×24.

STEP 1 Draw a model to show 15×24. Break apart the factors into tens and ones to show the partial products.	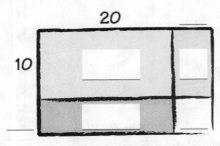

STEP 2 Write the product for each of the smaller rectangles.	(10 × 2 tens) (10 × 4 ones) (5 × 2 tens) (5 × 4 ones) (10 × 20) (10 × 4) (5 × 20) (5 × 4) 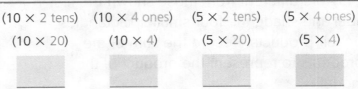

STEP 3 Add to find the product for the whole model.	☐ + ☐ + ☐ + ☐ = ____

So, $15 \times 24 = 360$.

The model shows four parts. Each part represents a partial product. The partial products are 200, 40, 100, and 20.

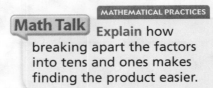

Math Talk MATHEMATICAL PRACTICES
Explain how breaking apart the factors into tens and ones makes finding the product easier.

© Houghton Mifflin Harcourt Publishing Company

Multiply Using Partial Products

Essential Question How can you use place value and partial products to multiply 2-digit numbers?

 UNLOCK the Problem REAL WORLD

CONNECT You know how to break apart a model to find partial products. How can you use what you know to find and record a product?

 Multiply. 34 × 57 **Estimate.** 30 × 60 = _____

SHADE THE MODEL	THINK AND RECORD

STEP 1

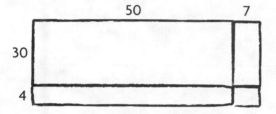

$$\begin{array}{r} 57 \\ \times\,34 \\ \hline \end{array}$$

← Multiply the tens by the tens.
30 × 5 tens = 150 tens

STEP 2

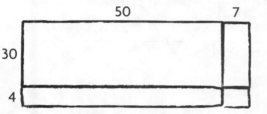

$$\begin{array}{r} 57 \\ \times\,34 \\ \hline 1{,}500 \end{array}$$

← Multiply the ones by the tens.
30 × 7 ones = 210 ones

STEP 3

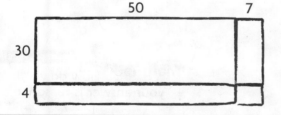

$$\begin{array}{r} 57 \\ \times\,34 \\ \hline 1{,}500 \\ 210 \end{array}$$

← Multiply the tens by the ones.
4 × 5 tens = 20 tens

STEP 4

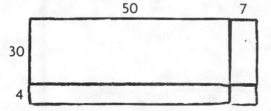

$$\begin{array}{r} 57 \\ \times\,34 \\ \hline 1{,}500 \\ 210 \\ 200 \\ +\, \end{array}$$

← Multiply the ones by the ones.
4 × 7 ones = 28 ones
← Add the partial products.

So, 34 × 57 = 1,938. Since 1,938 is close to the estimate of 1,800, it is reasonable.

Math Talk MATHEMATICAL PRACTICES
You can write 10 × 4 ones = 40 ones as 10 × 4 = 40. What is another way to write 10 × 3 tens = 30 tens?

🔒 Example

The apples from each tree in an orchard can fill 23 bushel baskets. If 1 row of the orchard has 48 trees, how many baskets of apples can be filled?

Multiply. 48 × 23   **Estimate.** 50 × 20 = _____

	THINK	RECORD

STEP 1

Multiply the tens by the tens.

23
× 48
_____

← 40 × _____ tens = _____ tens

STEP 2

Multiply the ones by the tens.

23
× 48
_____
800

← 40 × _____ ones = _____ ones

STEP 3

Multiply the tens by the ones.

23
× 48
_____
800
120

← 8 × _____ tens = _____ tens

STEP 4

Multiply the ones by the ones. Then add the partial products.

23
× 48
_____
800
120
160
+ _____

← 8 × _____ ones = _____ ones

So, 1,104 baskets can be filled.

Math Talk MATHEMATICAL PRACTICES
How do you know your answer is reasonable?

Share and Show 📝 MATH BOARD

1. Find 24 × 34.

	30	4
20	600	80
4	120	16

```
    3 4
  × 2 4
```

Name _____

Record the product.

2. 12
    × 12
    ‾‾‾‾

3. 31
    × 24
    ‾‾‾‾

✓ 4. 25
    × 43
    ‾‾‾‾

✓ 5. 37
    × 26
    ‾‾‾‾

MATHEMATICAL PRACTICES

Math Talk Explain how to model and record 74 × 25.

On Your Own

Record the product.

6. 54
    × 15
    ‾‾‾‾

7. 87
    × 16
    ‾‾‾‾

8. 62
    × 56
    ‾‾‾‾

9. 49
    × 63
    ‾‾‾‾

Practice: Copy and Solve Record the product.

10. 38 × 47

11. 46 × 27

12. 72 × 53

13. 98 × 69

14. 53 × 68

15. 76 × 84

16. 92 × 48

17. 37 × 79

 Algebra Find the unknown digits. Complete the problem.

18. ▦ 6
    × ▦ 4
    ‾‾‾‾‾‾
     1,400
       120
       280
    +   24
    ‾‾‾‾‾‾

19. ▦ 2
    × ▦ 7
    ‾‾‾‾‾‾
     7,200
       180
       560
    +   14
    ‾‾‾‾‾‾

20. ▦ 6
    × ▦ 5
    ‾‾‾‾‾‾
     1,500
       300
        90
    +   18
    ‾‾‾‾‾‾

21. 3 ▦
    × ▦ 8
    ‾‾‾‾‾‾
       600
        80
       240
    +   32
    ‾‾‾‾‾‾

Problem Solving REAL WORLD

Use the pictograph for 22–24.

22. A fruit-packing warehouse is shipping 15 boxes of grapefruit to a store in St. Louis, Missouri. What is the total weight of the shipment?

_____

23. How much less do 13 boxes of tangelos weigh than 18 boxes of tangerines?

_____

24. What is the weight of 12 boxes of oranges?

_____

Pounds of Citrus Fruit per Box	
Citrus Fruit	**Weight per Box (in pounds)**
Grapefruit	⚬ ⚬ ⚬ ⚬ ⚬ ⚬ ⚬ ◖
Orange	⚬ ⚬ ◖ ⚬ ⚬ ⚬ ◖ ⚬ ⚬
Tangelo	⚬ ⚬ ◖ ⚬ ⚬ ⚬ ◖ ⚬ ⚬
Tangerine	⚬ ⚬ ⚬ ⚬ ⚬ ⚬ ⚬ ⚬ ◖

Key: Each ⚬ = 10 pounds.

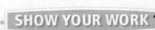

SHOW YOUR WORK

25. **H.O.T.** Each person in the United States eats about 65 fresh apples each year. Based on this estimate, how many apples do 3 families of 4 eat each year?

_____

26. **Write Math** ➤ The product 26 × 93 is more than 25 × 93. How much more? **Explain** how you know without multiplying.

_____

_____

_____

27. **Test Prep** Each row of apple trees has 14 trees. There are 16 rows. How many apple trees are there?

   (A) 1,340   (C) 184

   (B) 224   (D) 124

FOR MORE PRACTICE:
Standards Practice Book, pp. P57–P58

Name _____

 Mid-Chapter Checkpoint

▶ **Concepts and Skills**

1. Explain how to find 40×50 using mental math.

   _____

   _____

2. What is the first step in estimating 56×27?

   _____

Choose a method. Then find the product.

3. 35×10 _____

4. 19×20 _____

5. 12×80 _____

6. 70×50 _____

7. 58×40 _____

8. 30×40 _____

9. 14×60 _____

10. 20×30 _____

11. 16×90 _____

Estimate the product. Choose a method.

12. 81×38 _____

13. $16 \times \$59$ _____

14. 43×25 _____

15. 76×45 _____

16. $65 \times \$79$ _____

17. 92×38 _____

18. 37×31 _____

19. $26 \times \$59$ _____

20. 54×26 _____

21. 52×87 _____

22. 39×27 _____

23. 63×58 _____

Fill in the bubble completely to show your answer.

24. Ms. Traynor's class is taking a field trip to the zoo. The trip will cost $26 for each student. There are 22 students in her class. Which is the best estimate for the cost of the students' field trip?

Ⓐ $480

Ⓑ $600

Ⓒ $1,200

Ⓓ $6,000

25. Tito wrote the following on the board. What is the unknown number?

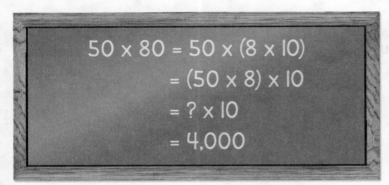

$$50 \times 80 = 50 \times (8 \times 10)$$
$$= (50 \times 8) \times 10$$
$$= ? \times 10$$
$$= 4,000$$

Ⓐ   40

Ⓑ   58

Ⓒ   400

Ⓓ 4,000

26. Which shows a way to find 15×32?

Ⓐ $(10 \times 3) + (10 \times 2) + (30 \times 1) + (30 \times 50)$

Ⓑ $(10 \times 30) + (10 \times 2) + (50 \times 30) + (50 \times 2)$

Ⓒ $(10 + 30) + (10 + 2) + (30 + 10) + (30 + 5)$

Ⓓ $(10 \times 30) + (10 \times 2) + (5 \times 30) + (5 \times 2)$

27. The cost of a ski-lift ticket is $31. How much will 17 tickets cost?

Ⓐ $48

Ⓒ $310

Ⓑ $217

Ⓓ $527

Name _____

Multiply with Regrouping

Essential Question How can you use regrouping to multiply 2-digit numbers?

 UNLOCK the Problem REAL WORLD

By 1914, Henry Ford had streamlined his assembly line to make a Model T Ford car in 93 minutes. How many minutes did it take to make 25 Model Ts?

▲ The first production Model T Ford was assembled on October 1, 1908.

🔑 **Use place value and regrouping.**

Multiply. 93 × 25   **Estimate.** 90 × 30 = _____

	THINK	RECORD
STEP 1	• Think of 93 as 9 tens and 3 ones. • Multiply 25 by 3 ones.	1 25 × 93 _____ ← 3 × 25
STEP 2	• Multiply 25 by 9 tens.	4 1̸ 25 × 93 ‾‾‾‾ 75 _____ ← 90 × 25
STEP 3	• Add the partial products.	4 1̸ 25 × 93 ‾‾‾‾ 75 2,250 _____

So, 93 × 25 is 2,325. Since _____ is close

to the estimate of _____, the answer is reasonable.

Math Talk   MATHEMATICAL PRACTICES
Explain why you will get the same answer whether you multiply 93 × 25 or 25 × 93.

Different Ways to Multiply You can use different ways to multiply and still get the correct answer. Shawn and Patty both solved 67×40 correctly, but they used different ways.

Look at Shawn's paper.

60	x	40	=	2,400
7	x	40	=	280
2,400	+	280	=	2,680

So, Shawn's answer is $67 \times 40 = 2,680$.

Look at Patty's paper.

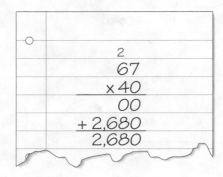

```
   2
   67
 × 40
   00
+2,680
 2,680
```

So, Patty also found $67 \times 40 = 2,680$.

1. What method did Shawn use to solve the problem?

_____

2. What method did Patty use to solve the problem?

_____

Share and Show . . . .

1. Look at the problem. Complete the sentences.

Multiply _____ and _____ to get 0.

Multiply _____ and _____ to get 1,620.

Add the partial products.

$0 + 1,620 =$ _____

$$\begin{array}{r} 4 \\ 27 \\ \times 60 \\ \hline 0 \\ +1,620 \\ \hline \end{array}$$

Name _____

Estimate. Then find the product.

2. Estimate: _____

$$
\begin{array}{r}
68 \\
\times\ 53 \\
\hline
\end{array}
$$

3. Estimate: _____

$$
\begin{array}{r}
61 \\
\times\ 54 \\
\hline
\end{array}
$$

4. Estimate: _____

$$
\begin{array}{r}
90 \\
\times\ 27 \\
\hline
\end{array}
$$

MATHEMATICAL PRACTICES

Math Talk Explain why you can omit zeros of the first partial product when you multiply 20 × 34.

On Your Own ....

Estimate. Then find the product.

5. Estimate: _____

$$
\begin{array}{r}
30 \\
\times\ 47 \\
\hline
\end{array}
$$

6. Estimate: _____

$$
\begin{array}{r}
78 \\
\times\ 56 \\
\hline
\end{array}
$$

7. Estimate: _____

$$
\begin{array}{r}
27 \\
\times\ 25 \\
\hline
\end{array}
$$

Practice: Copy and Solve Estimate. Then find the product.

8. 34×65

9. $42 \times \$13$

10. 60×17

11. 62×45

12. $57 \times \$98$

13. $92 \times \$54$

14. 75×20

15. 66×55

16. $73 \times \$68$

17. 72×40

H.O.T. **Algebra** Write a rule for the pattern. Use your rule to find the unknown numbers.

18.

Hours	h	5	10	15	20	25
Minutes	m	300	600	900		

Rule: _____

19.

Minutes	m	12	14	16	18	20
Seconds	s	720	840		1,080	

Rule: _____

🔑 UNLOCK the Problem > REAL WORLD

20. Machine A can label 11 bottles in 1 minute.
Machine B can label 12 bottles in 1 minute.
How many bottles can both machines label in 15 minutes?

(A) 165   (C) 245

(B) 180   (D) 345

a. What do you need to know? _____

_____

b. What numbers will you use? _____

c. Tell why you might use more than one operation to solve the problem.

_____

_____

_____

d. Solve the problem.

e. Fill in the bubble for the correct answer choice above.

21. A toy company makes wooden blocks. A carton holds 85 blocks. How many blocks can 19 cartons hold?

(A) 1,615

(B) 1,575

(C) 1,515

(D) 850

22. A company is packing cartons of candles. Each carton can hold 75 candles. If 50 cartons are packed, how many candles have been packed?

(A) 375

(B) 3,500

(C) 3,550

(D) 3,750

Name _____

Choose a Multiplication Method

Essential Question How can you find and record products of two 2-digit numbers?

🔑 UNLOCK the Problem REAL WORLD

Did you know using math can help prevent you from getting a sunburn?

The time it takes to burn without sunscreen multiplied by the SPF, or sun protection factor, is the time you can stay in the sun safely with sunscreen.

If today's UV index is 8, Erin will burn in 15 minutes without sunscreen. If Erin puts on lotion with an SPF of 25, how long will she be protected?

- Underline the sentence that tells you how to find the answer.
- Circle the numbers you need to use. What operation will you use?

_____

▲ Sunscreen helps to prevent sunburn.

🔓 One Way Use partial products to find 15 × 25.

```
        25
      × 15
   _____
   _____    ← 10 × 2 tens  =  20 tens
   _____    ← 10 × 5 ones  =  50 ones
   _____    ←  5 × 2 tens  =  10 tens
 + _____    ←  5 × 5 ones  =  25 ones
   _____    ← Add.
```

✏️ **Draw a picture to check your work.**

Math Talk MATHEMATICAL PRACTICES The product is 375. Explain what 375 means for Erin.

🔑 Another Way Use regrouping to find 15 × 25.

Estimate. 20 × 20 = _____

STEP 1

Think of 15 as 1 ten 5 ones.
Multiply 25 by 5 ones, or 5.

$$\begin{array}{r} \overset{2}{25} \\ \times\ 15 \\ \hline \end{array}$$ ← 5 × 25

STEP 2

Multiply 25 by 1 ten, or 10.

$$\begin{array}{r} \overset{2}{25} \\ \times\ 15 \\ \hline 125 \end{array}$$ ← 10 × 25

STEP 3

Add the partial products.

$$\begin{array}{r} \overset{2}{25} \\ \times\ 15 \\ \hline 125 \\ +\ 250 \\ \hline \end{array}$$

Try This! Multiply. 57 × $43

Estimate. 57 × $43

Use partial products.

$$\begin{array}{r} \$\ 4\ 3 \\ \times\quad 5\ 7 \\ \hline \end{array}$$

Use regrouping.

$$\begin{array}{r} \$\ 4\ 3 \\ \times\quad 5\ 7 \\ \hline \end{array}$$

1. How do you know your answer is reasonable?

_____

_____

2. Look at the partial products and regrouping methods above.
How are the partial products 2,000 and 150 related to 2,150?

_____

How are the partial products 280 and 21 related to 301?

_____

Name _____

Share and Show

1. Find the product.

		5	4
×		2	9

Math Talk MATHEMATICAL PRACTICES
Explain why you begin with the ones place when you use the regrouping method to multiply.

Estimate. Then choose a method to find the product.

2. Estimate: _____
$$\begin{array}{r} 36 \\ \times\ 14 \\ \hline \end{array}$$

3. Estimate: _____
$$\begin{array}{r} 63 \\ \times\ 42 \\ \hline \end{array}$$

4. Estimate: _____
$$\begin{array}{r} 84 \\ \times\ 53 \\ \hline \end{array}$$

5. Estimate: _____
$$\begin{array}{r} 71 \\ \times\ 13 \\ \hline \end{array}$$

On Your Own

Estimate. Then choose a method to find the product.

6. Estimate: _____
$$\begin{array}{r} 34 \\ \times\ 48 \\ \hline \end{array}$$

7. Estimate: _____
$$\begin{array}{r} 19 \\ \times\ 41 \\ \hline \end{array}$$

8. Estimate: _____
$$\begin{array}{r} \$33 \\ \times\ 17 \\ \hline \end{array}$$

9. Estimate: _____
$$\begin{array}{r} 28 \\ \times\ 39 \\ \hline \end{array}$$

Practice: Copy and Solve Estimate. Find the product.

10. $29 \times \$82$

11. 57×79

12. 80×27

13. $32 \times \$75$

14. 55×48

15. $19 \times \$82$

16. $25 \times \$25$

17. 41×98

H.O.T. **Algebra** Use mental math to find the number.

18. $30 \times 14 = 420$, so $30 \times 15 =$ _____.

19. $25 \times 12 = 300$, so $25 \times$ _____ $= 350$.

🔑 UNLOCK the Problem › REAL WORLD

20. Martin collects stamps. He counted 48 pages in his collector's album. The first 20 pages each have 35 stamps in 5 rows. The rest of the pages each have 54 stamps. How many stamps does Martin have in his album?

a. What do you need to know? _____

_____

b. How will you use multiplication to find the number of stamps? _____

_____

c. Tell why you might use addition and subtraction to help solve the problem.

_____

_____

d. Show the steps to solve the problem.

e. Complete the sentences.

Martin has a total of _____ stamps on the first 20 pages.

There are _____ more pages after the first 20 pages in Martin's album.

There are _____ stamps on the rest of the pages.

There are _____ stamps in the album.

21. Each of the 25 students in a group read for 45 minutes. How many minutes did the group spend reading?

_____

22. Test Prep Each row of peach trees has 37 trees. There are 16 rows. How many peach trees are there?

Ⓐ 53

Ⓑ 259

Ⓒ 342

Ⓓ 592

Name _____

Problem Solving • Multiply 2-Digit Numbers

Essential Question How can you use the strategy *draw a diagram* to solve multistep multiplication problems?

🔑 UNLOCK the Problem REAL WORLD

During the 2010 Great Backyard Bird Count, an average of 42 bald eagles were counted in each of 20 locations throughout Alaska. In 2009, an average of 32 bald eagles were counted in each of 26 locations throughout Alaska. Based on this data, how many more bald eagles were counted in 2010 than in 2009?

Use the graphic organizer to help you solve the problem.

Read the Problem	Solve the Problem
What do I need to find? I need to find _____ bald eagles were counted in 2010 than in 2009.	• First, find the total number of bald eagles counted in 2010. _____ × _____ = _____ bald eagles counted in 2010
What information do I need to use? In 2010, _____ locations counted an average of _____ bald eagles each. In 2009 _____ locations counted an average of _____ bald eagles each.	• Next, find the total number of bald eagles counted in 2009. = _____ × _____ = _____ bald eagles counted in 2009
How will I use the information? I can solve simpler problems. Find the number of bald eagles counted in _____. Find the number of bald eagles counted in _____. Then draw a bar model to compare the _____ count to the _____ count.	• Last, draw a bar model. I need to subtract. ┌──────────────────────────┐ │ 840 bald eagles in 2010 │ └──────────────────────────┘ ┌─────────────────────┐ │ 832 bald eagles in 2009 │──┐ └─────────────────────┘ ? 840 − 832 = _____ So, there were _____ more bald eagles counted in 2010 than in 2009.

🔑 Try Another Problem

Prescott Valley, Arizona, reported a total of 29 mourning doves in the Great Backyard Bird Count. Mesa, Arizona, reported 20 times as many mourning doves as Prescott Valley. If Chandler reported a total of 760 mourning doves, how many more mourning doves were reported in Chandler than in Mesa?

Mourning dove ▲

Read the Problem	Solve the Problem
What do I need to find?	
What information do I need to use?	
How will I use the information?	760 mourning doves in Chandler
	580 mourning doves in Mesa ⌐?⌐

- Is your answer reasonable? **Explain.** _____

_____

_____

Math Talk Describe another way you could solve this problem.

Name _____

Share and Show

UNLOCK the Problem **Tips**

√ Underline important facts.
√ Choose a strategy.
√ Use the Problem Solving MathBoard.

1. An average of 74 reports with bird counts were turned in each day in June. An average of 89 were turned in each day in July. How many reports were turned in for both months? (Hint: There are 30 days in June and 31 days in July.)

   First, write the problem for June.

   _____

   Next, write the problem for July.

   _____

   Last, find and add the two products.

   _____ reports were turned in for both months.

2. **H.O.T.** **What if** an average of 98 reports were turned in each day for the month of June? How many reports were turned in for June? **Describe** how your answer for June would be different.

   _____

   _____

   _____

3. On each of Maggie's bird-watching trips, she has seen at least 24 birds. If she has taken 4 of these trips each year over the past 16 years, at least how many birds has Maggie seen?

   _____

4. Each of 5 bird-watchers reported seeing 15 roseate spoonbills in a day. If they each reported seeing the same number of roseate spoonbills over 14 days, how many would be reported?

   _____

SHOW YOUR WORK

© Houghton Mifflin Harcourt Publishing Company

On Your Own......

Choose a STRATEGY

Act It Out
Draw a Diagram
Find a Pattern
Make a Table or List
Solve a Simpler Problem

5. **H.O.T.** There are 12 inches in a foot. In September, Mrs. Harris orders 32 feet of ribbon for the Crafts Club. In January, she orders 9 fewer feet. How many inches of ribbon does Mrs. Harris order? **Explain** how you found your answer.

_____

_____

_____

6. Lydia is having a party on Saturday. She decides to write a riddle on her invitations to describe her house number on Cypress Street. Use the clues to find Lydia's address.

_____

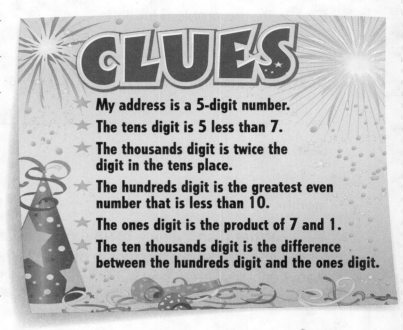

CLUES

★ My address is a 5-digit number.
★ The tens digit is 5 less than 7.
★ The thousands digit is twice the digit in the tens place.
★ The hundreds digit is the greatest even number that is less than 10.
★ The ones digit is the product of 7 and 1.
★ The ten thousands digit is the difference between the hundreds digit and the ones digit.

7. Nationwide, participants in the 2008 Great Backyard Bird Count reported seeing 778,871 Canada geese and 363,321 American crows. How many more Canada geese were seen than American crows?

SHOW YOUR WORK

_____

8. **Test Prep** Carol is the treasurer of her bird-watching club. The club wants to order shirts for each of the 18 members. If each shirt costs $21, what is the cost for the members' shirts?

  (A) $39   (C) $380

  (B) $378   (D) $540

Name _____

 **Chapter Review/Test**

▶ **Concepts and Skills**

1. Explain how to find 14 × 19 by breaking apart the factors into tens and ones and finding the sum of the four partial products.

_____

_____

2. Explain how to find 40 × 80 using mental math.

_____

Estimate the product. Choose a method.

3. 80 × 26   4. 19 × $67   5. 43 × 25   6. 54 × 83

_____   _____   _____   _____

Estimate. Then find the product.

7. Estimate:_____

$24
× 96

8. Estimate:_____

44
× 60

9. Estimate:_____

99
× 14

10. Estimate:_____

67
× 25

11. Estimate:_____

36
× 57

12. Estimate:_____

$54
× 29

13. Estimate:_____

76
× 38

14. Estimate:_____

85
× 46

Fill in the bubble completely to show your answer.

15. Each month Sid's parents put $75 into his college fund. How much do his parents put in the fund during 2 years?

  (A) $150

  (B) $450

  (C) $1,800

  (D) $15,300

16. Mrs. Jenks wrote the correct answer to a homework problem on the board below. Which of the following could have been the homework problem?

  (A) 5 × 4,000

  (B) 50 × 400

  (C) 50 × 40

  (D) 50 × 4,000

17. George buys 30 cartons of 18 eggs for the Community Pancake Breakfast. How many eggs does he buy?

  (A) 340

  (B) 354

  (C) 460

  (D) 540

Name _____

Fill in the bubble completely to show your answer.

18. Mrs. Sampson donated a carton of pencils for each of the 35 classes at Lancet Elementary School. Each carton holds 64 pencils. Which is the best estimate for the number of pencils Mrs. Sampson donated?

Ⓐ 99

Ⓑ 1,800

Ⓒ 2,400

Ⓓ 2,800

19. The school's athletic department ordered 95 dozen badminton feather shuttles. How many feather shuttles were ordered?

Ⓐ 2,280   Ⓒ 1,030

Ⓑ 1,140   Ⓓ 114

20. Jill sold 35 adult tickets and 48 child tickets for a fund-raising dinner. An adult ticket costs $18 and a child ticket costs $14. How much did Jill collect for the tickets?

Ⓐ $1,354   Ⓒ $1,232

Ⓑ $1,302   Ⓓ $1,102

21. Which shows a way to find 35×74?

Ⓐ $(30 \times 7) + (30 \times 4) + (70 \times 3) + (70 \times 5)$

Ⓑ $(30 \times 70) + (30 \times 4) + (50 \times 70) + (50 \times 4)$

Ⓒ $(30 + 70) + (30 + 4) + (70 + 30) + (70 + 5)$

Ⓓ $(30 \times 70) + (30 \times 4) + (5 \times 70) + (5 \times 4)$

22. New seats are being delivered to the theater. There are 45 new seats for each row in a 15-row section. How many seats are being delivered?

Ⓐ 60   Ⓒ 675

Ⓑ 400   Ⓓ 1,000

23. Gulfside Gifts has 48 boxes of postcards to sell. There are 24 postcards in each box. If the shop sells 3 boxes of postcards, how many postcards does the shop have left to sell? Explain how you found the answer.

_____

_____

24. Several steps in finding the product of 68 and 34 are shown below. Describe the remaining steps. Use pictures, words, or numbers. Then complete the multiplication.

$$
\begin{array}{r}
\overset{2}{\overset{3}{6}}8 \\
\times\ 34 \\
\hline
272 \\
40 \\
\end{array}
$$

_____

_____

▶ **Performance Task**

25. A city is having a festival in a local park. Alison's Bakery has agreed to donate $1,200 worth of baked goods for the event. The city wants to order 12 loaves of holiday bread, 18 dozen biscuits, 12 dozen bagels, and 14 dozen multigrain rolls.

Ⓐ Is the cost of the baked goods under the $1,200 donation limit? Use pictures, numbers, or words to explain how you found your answer.

_____

_____

Ⓑ If yes, what could the city add to the order? If no, what could the city remove from the order?

_____

_____

Price List	
Baked Goods	**Group Size**
Holiday Bread	$20
Biscuits	$12/dozen
Bagels	$28/dozen
Multigrain Rolls	$22/dozen

Chapter 4

Divide by 1-Digit Numbers

Show What You Know

Check your understanding of important skills.

Name _____

▶ **Use Arrays to Divide** Draw to complete each array.
Then complete the number sentence.

1. ■ ■ ■ ■

  8 ÷ 4 = _____

2. ■
  ■
  ■

  21 ÷ 3 = _____

▶ **Multiples** Write the first six multiples of the number.

3. 4: _____

4. 10: _____

▶ **Subtract Through 4-Digit Numbers** Find the difference.

5. 626
   −   8

6. 744
   − 36

7. 5,413
   −2,037

8. 8,681
   − 422

MATH DETECTIVE
WITH
CARMEN SANDIEGO™

Each digit in the division example has
been replaced with the same letter
throughout. (r stands for remainder.)
The digits used were 1, 2, 3, 4, 5, 7, and 9.
Be a Math Detective and find the
numbers. Clue: U is 5.

```
      SU rE
  U)CAN
    −CU
     IN
    −IU
      E
```

© Houghton Mifflin Harcourt Publishing Company

Vocabulary Builder

▶ **Visualize It** •••••••••••••••••••••••••••••••••••

Sort the words into the Venn diagram.

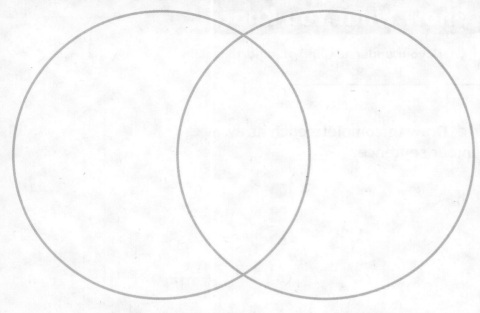

Multiplication Words   **Division Words**

Review Words

Distributive Property

divide

dividend

division

divisor

factor

multiplication

product

quotient

Preview Words

compatible numbers

multiple

partial quotient

remainder

▶ **Understand Vocabulary** •••••••••••••••••••••••

Write the word that answers the riddle.

1. I am the method of dividing in which multiples of the divisor are subtracted from the dividend and then the quotients are added together.

   _____

2. I am the number that is to be divided in a division problem.

   _____

3. I am the amount left over when a number cannot be

   divided equally. _____

4. I am the number that divided the dividend.

   _____

Name _____

Estimate Quotients Using Multiples

Essential Question How can you use multiples to estimate quotients?

🔑 UNLOCK the Problem REAL WORLD

The bakery made 110 pumpkin muffins. They will be packed in boxes with 8 muffins in each box. About how many boxes will there be?

You can use multiples to estimate.

A **multiple** of a number is the product of a number and a counting number. 1, 2, 3, 4, and so on, are counting numbers.

🔑 **Estimate. 110 ÷ 8**

Think: What number multiplied by 8 is about 110?

STEP 1 List the multiples of 8 until you reach 110 or greater.

Counting number	1	2	3	4	5	6	7	8	9	10	11	12	13	14
Multiple of 8	8	16	24	32			56	64				96		112

STEP 2 Find the multiples of 8 that 110 is between.

13 × 8 = _____

14 × 8 = _____

110 is between _____ and _____, so 110 ÷ 8 is between 13 and 14.

110 is closest to _____, so 110 ÷ 8 is about _____.

So, there will be about _____ boxes.

Try This!

List the next 8 multiples of 10.

10, 20, _____

List the next 7 multiples of 100.

100, 200, _____

Math Talk
MATHEMATICAL PRACTICES
When estimating a quotient, how do you know which two numbers it is between? **Explain.**

🔑 Example Estimate 196 ÷ 4

Think: What number times 4 is about 196?

STEP 1 List the next 6 multiples of 4.

4, 8, 12, 16, _____

Are any multiples close to 196? _____
_____

Think: If I multiply by multiples of 10, the products will be greater. Using multiples of 10 will get me to 196 faster.

STEP 2 Multiply 4 by multiples of 10.

$10 \times 4 = 40$

$20 \times 4 = 80$

$30 \times 4 =$ _____

$40 \times 4 =$ _____

$50 \times 4 =$ _____

The quotient is between 40 and 50.

_____ $\times$ 4 is closest to _____, so 196 ÷ 4 is about _____.

Share and Show ·

1. A restaurant has 68 chairs. There are six chairs at each table. About how many tables are in the restaurant?

Estimate. 68 ÷ 6

   Think: What number times 6 is about 68?

   $10 \times 6 =$ _____

   $11 \times 6 =$ _____

   $12 \times 6 =$ _____

   68 is closest to _____, so the best estimate is

   about _____ tables are in the restaurant.

MATHEMATICAL PRACTICES

Math Talk When do you multiply the divisor by multiples of 10 to estimate a quotient? **Explain.**

© Houghton Mifflin Harcourt Publishing Company

138

Name _____

Find two numbers the quotient is between. Then estimate the quotient.

✅ **2.** $41 \div 3$

_____

_____

✅ **3.** $192 \div 5$

_____

_____

On Your Own ....

Find two numbers the quotient is between. Then estimate the quotient.

4. $90 \div 7$

_____

_____

5. $67 \div 4$

_____

_____

6. $281 \div 9$

_____

_____

7. $102 \div 7$

_____

_____

8. $85 \div 6$

_____

_____

9. $220 \div 8$

_____

_____

10. $443 \div 5$

_____

_____

11. $95 \div 8$

_____

_____

12. $49 \div 3$

_____

_____

13. $249 \div 8$

_____

_____

14. $412 \div 7$

_____

_____

15. $177 \div 9$

_____

_____

H.O.T. **Decide whether the actual quotient is greater than or less than the estimate given. Write < or >.**

16. $83 \div 8 \bigcirc 10$

17. $155 \div 4 \bigcirc 40$

18. $70 \div 6 \bigcirc 11$

19. $416 \div 5 \bigcirc 80$

20. $194 \div 2 \bigcirc 90$

21. $200 \div 3 \bigcirc 70$

Problem Solving **REAL WORLD**

22. **H.O.T.** If a bottlenose dolphin can eat 175 pounds of fish, squid, and shrimp in a week, about how many pounds of food does it eat in a day? Milo says the answer is about 20 pounds. Leah says the answer is about 30 pounds. Who is correct? **Explain.**

_____

_____

_____

_____

23. A mother bottlenose ate about 278 pounds of food in one week. About how much food did she eat in a day?

_____

SHOW YOUR WORK

24. **Write Math** ➤ Four families went out for lunch. The total food bill came to $167. The families also left a $30 tip for the waitress. If each family spent the same amount, about how much did each family spend on dinner? **Explain** how you found your answer.

_____

_____

_____

25. **What's the Question?** A dolphin's heart beats 688 times in 6 minutes. Answer: about 100 times.

_____

_____

26. **Test Prep** Small groups of about 7 bottlenose dolphins live together in pods. Sometimes several pods join in a herd to help protect each other. About how many pods are there in a herd of 204 dolphins?

   Ⓐ about 20   Ⓒ about 40

   Ⓑ about 30   Ⓓ about 1,400

Remainders

Essential Question How can you use models to divide whole numbers that do not divide evenly?

Investigate

Materials ■ counters

Erica and 2 friends are playing a game of dominoes. There are 28 dominoes in the set. Erica wants each player to receive the same number of dominoes. Can she divide them equally among the 3 players? Why or why not?

You can use division to find the number of dominoes each player will receive.

A. Use 28 counters to represent the 28 dominoes. Then draw 3 circles to represent the 3 players.

B. Share the counters equally among the 3 groups by placing them in the circles.

> **Draw a quick picture to show your work.**

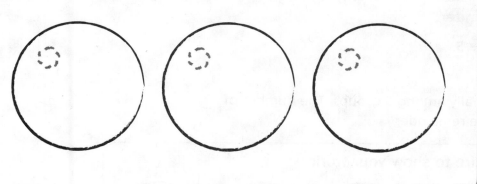

C. Find the number of counters in each group and the number of counters left over. Record your answer.

_____ counters in each group

_____ counter left over

Draw Conclusions

1. How many dominoes does each player receive? _____

How many dominoes are left over? _____

2. **H.O.T.** **Explain** how the model helped you find the number of dominoes each player receives. Why is 1 counter left outside the equal groups?

_____

_____

_____

3. **Apply** Use counters to represent a set of 28 dominoes. How many players can play dominoes if each player receives 9 dominoes? Will any dominoes be left over? Explain.

_____

_____

Make Connections

When a number cannot be divided evenly, the amount left over is called the **remainder**.

Use counters to find 39 ÷ 5.

• Use 39 counters.

• Share the counters equally among 5 groups. The number of counters left over is the remainder.

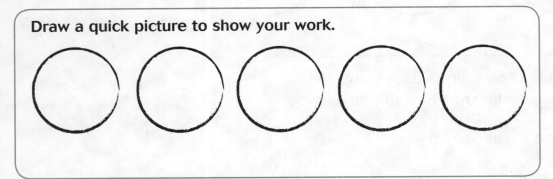

Draw a quick picture to show your work.

For 39 ÷ 5, the quotient is _____ and the remainder

is _____, or 7 r4.

Math Talk
© Houghton Mifflin Harcourt Publishing Company

MATHEMATICAL PRACTICES
How do you know when there will be a remainder in a division problem?

Name _____

Share and Show   ●

Use counters to find the quotient and remainder.

1. $10 \div 3$　　　　**2.** $28 \div 5$　　　　**3.** $15 \div 6$　　　　**4.** $11 \div 3$

_____　　_____　　_____　　_____

5. $9\overline{)26}$　　　　**6.** $22 \div 3$　　　　**7.** $4\overline{)19}$　　　　**8.** $4\overline{)38}$

_____　　_____　　_____　　_____

9. $29 \div 4$　　　　**10.** $34 \div 5$　　　　**11.** $25 \div 3$　　　　✓**12.** $7\overline{)20}$

_____　　_____　　_____　　_____

Divide. Draw a quick picture to help.

13. $19 \div 3$

14. $5\overline{)47}$

_____

15. $4\overline{)35}$

✓**16.** $23 \div 8$

_____

17. **Write Math** ► **Explain** how you use a quick picture
to find the quotient and remainder.

_____

_____

_____

Problem Solving REAL WORLD

H.O.T. What's the Error?

18. Macy, Kayley, Maddie, and Rachel collected 13 marbles.
They want to share the marbles equally. How many
marbles will each of the 4 girls get? How many marbles
will be left over?

Frank used a model to solve this problem. He says his
model represents $4\overline{)13}$. What is his error?

**Look at the way Frank solved this
problem. Find and describe his error.**

_____

_____

_____

_____

_____

_____

_____

_____

_____

_____

_____

**Draw a correct model and solve
the problem.**

So, each of the 4 girls will get _____

marbles and _____ marble will be

left over.

Interpret the Remainder

Essential Question How can you use remainders in division problems?

> ### 🔑 UNLOCK the Problem > REAL WORLD

Magda has some leftover wallpaper 73 inches long. She wants to cut it into 8 pieces to use around the photos in her scrapbook. Each piece will have equal length. How long will each piece be?

When you solve a division problem with a remainder, the way you interpret the remainder depends on the situation and the question.

🔒 **One Way** Write the remainder as a fraction.

The divisor is _____ pieces.

The _____ is 73 inches.

Divide to find the quotient and remainder.

$$8\overline{)73} \quad \overset{9}{} \ r1$$

The remainder represents 1 inch left over, which can also be divided into 8 equal parts and written as a fraction.

$$\frac{\text{remainder}}{\text{divisor}} = \underline{\qquad}$$

> **Remember**
> You can use multiples, counters, or draw a quick picture to divide.

Write the quotient with the remainder written as a fraction. _____

So, each piece will be _____ inches long.

Try This!

Jim made 32 ounces of soup for 5 people. How many ounces will each person get? Complete the division.

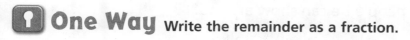

$$5\overline{)32} \quad \underline{}$$

Math Talk MATHEMATICAL PRACTICES

Explain what the 2 in the answer represents.

Each person gets _____ ounces.

🔑 Other Ways

Ⓐ Use only the quotient.

Ben is a tour guide at a glass-blowing studio. He can take no more than 7 people at a time on a tour. If 80 people want to see the glass-blowing demonstration, how many groups of 7 people will Ben show around?

First, divide to find the quotient and remainder.
Then, decide how to use the quotient and remainder.

The quotient is _____ .

$$7\overline{)80} \quad 11 \text{ r}$$

The remainder is _____ .

Ben can give tours to 7 people at a time. The quotient is the number of tour groups of exactly 7 people he can show around.

So, Ben gives tours to _____ groups of 7 people.

Ⓑ Add 1 to the quotient.

If Ben gives tours to all 80 people, how many tours will he give? A tour can have no more than 7 people. To show all 80 people around, Ben will have to give 1 more tour.

So, Ben will give _____ tours in all for 80 people.

Ⓒ Use only the remainder.

Ben gives tours to all 80 people. After he completes the tours for groups of 7 people, how many people are in his last tour?

The remainder is 3.

So, Ben's last tour will have _____ people.

Try This!

Students are driven to soccer games in vans. Each van holds 9 students. How many vans are needed for 31 students?

Divide. 31 ÷ 9 _____

Since there are _____ students left over, _____ vans are needed to carry 31 students.

© Houghton Mifflin Harcourt Publishing Company

MATHEMATICAL PRACTICES

Math Talk Explain why you would not write the remainder as a fraction when you find the number of vans needed.

Name _____

Share and Show

1. Olivia baked 53 mini-loaves of banana bread to be sliced for
   snacks at a craft fair. She will place an equal number of loaves
   in 6 different locations. How many loaves will be at each
   location?

   a. Divide to find the quotient and remainder. $6\overline{)53}$ ⎤ r ⎤

   b. Decide how to use the quotient and remainder
      to answer the question.

      _____

Interpret the remainder to solve.

✓ 2. **What if** Olivia wants to put only whole
     loaves at each location? How many
     loaves will be at each location?

     _____

✓ 3. Ed carves 22 small wooden animals to
     sell at the craft fair. He displays them in
     rows with 4 animals in a row. How many
     animals will not be in equal rows?

     _____

On Your Own

Interpret the remainder to solve.

4. Myra has a 17-foot roll of crepe paper to
   make 8 streamers to decorate for a party.
   How long will each streamer be if she
   cuts the roll into equal pieces?

   _____

5. Juan has a piano recital next month.
   Last week he practiced for 15 hours in
   all. Each practice session is 2 hours long.
   How many full practice sessions does
   Juan complete?

   _____

6. A total of 25 students sign up to be hosts on Parent's Night.
   Teams of 3 students greet parents. How many students cannot
   be on a team? **Explain.**

   _____

   _____

   _____

Problem Solving REAL WORLD

Use the picture for 7–9.

7. Cho is making sock puppets just like the one in the picture. If she has 53 buttons, how many puppets can she make?

   _____

8. **H.O.T.** **Pose a Problem** Write a question about Cho and the sock puppets for which the answer is 3. **Explain** the answer.

   _____

   _____

   _____

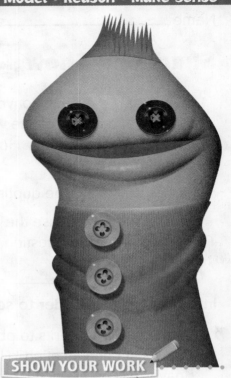

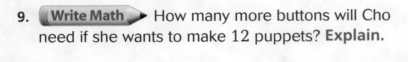

SHOW YOUR WORK

9. **Write Math** ▶ How many more buttons will Cho need if she wants to make 12 puppets? **Explain.**

   _____

   _____

   _____

10. **H.O.T.** Jonah cuts a board that is 33 inches long into 4 pieces of equal length. He uses the pieces for the sides of a picture frame. He puts an extra 2-inch wide trim on each side of the frame. How wide is the final frame?

    _____

11. **Test Prep** Mr. Alia gives a "Good Job" badge to each of the 74 students who help at a school event. There are 8 badges in a package. How many packages will he open?

    (A) 2   (C) $9\frac{1}{4}$

    (B) 9   (D) 10

FOR MORE PRACTICE:
Standards Practice Book, pp. P73–P74

Divide Tens, Hundreds, and Thousands

Essential Question How can you divide numbers through thousands by whole numbers through 10?

🔑 UNLOCK the Problem 〉 REAL 🌐 WORLD

Dustin is packing apples in gift boxes. Each gift box holds 4 apples. How many boxes can Dustin pack with 120 apples?

You can divide using basic facts and place value.

🔑 Example 1 Divide. 120 ÷ 4

STEP 1 Identify the basic fact.　　12 ÷ 4

STEP 2 Use place value.　　　　120 = _____ tens

STEP 3 Divide.　　　　　12 tens ÷ 4 = _____ tens　← **Think:** 4 × 3 tens = 12 tens

　　　　　　　　　　　　　　　= _____

　　　　　　　　　　120 ÷ 4 = 30

So, Dustin can pack _____ boxes.

🔑 Example 2 Divide. 1,200 ÷ 4

STEP 1 Identify the basic fact.　　12 ÷ 4

STEP 2 Use place value.　　　　1,200 = _____ hundreds

STEP 3 Divide.　　　12 hundreds ÷ 4 = _____ hundreds　← **Think:** 4 × 3 hundreds = 12 hundreds

　　　　　　　　　　　　　　= _____

　　　　　　　1,200 ÷ 4 = 300

Math Talk MATHEMATICAL PRACTICES
Describe the pattern in the place value of the dividends and quotients.

- **Explain** how to use a basic fact and place value to divide 4,000 ÷ 5.

_____

_____

Share and Show

1. Divide. 2,800 ÷ 7

What basic fact can you use? _____

2,800 = 28 _____

28 hundreds ÷ 7 = _____

2,800 ÷ 7 = _____

Math Talk MATHEMATICAL PRACTICES
Explain how Exercises 1 and 2 are alike and different.

2. Divide. 280 ÷ 7

What basic fact can you use? _____

280 = 28 _____

28 tens ÷ _____ = 4 _____

280 ÷ 7 = _____

Use basic facts and place value to find the quotient.

☑ **3.** 360 ÷ 6 = _____   **4.** 2,000 ÷ 5 = _____   ☑ **5.** 4,500 ÷ 9 = _____

On Your Own

Use basic facts and place value to find the quotient.

6. 560 ÷ 8 = _____   **7.** 200 ÷ 5 = _____   **8.** 240 ÷ 4 = _____

9. 810 ÷ 9 = _____   **10.** 6,400 ÷ 8 = _____   **11.** 3,500 ÷ 7 = _____

12. 5,000 ÷ 5 = _____   **13.** 9,000 ÷ 3 = _____   **14.** 3,000 ÷ 5 = _____

Algebra Find the unknown number.

15. 420 ÷ ■ = 60 _____   **16.** ■ ÷ 4 = 30 _____   **17.** 810 ÷ ■ = 90 _____

18. H.O.T. Divide 400 ÷ 40. **Explain** how patterns and place value can help.

_____

_____

_____

Name _____

Find the quotient.

19. $25 \div 5 =$ _____

$250 \div 5 =$ _____

$2{,}500 \div 5 =$ _____

20. $18 \div 2 =$ _____

$180 \div 2 =$ _____

$1{,}800 \div 2 =$ _____

21. $63 \div 9 =$ _____

$630 \div 9 =$ _____

$6{,}300 \div 9 =$ _____

22. Explain what you notice about the quotients in Exercises 19–21.

_____

_____

Problem Solving  REAL WORLD

23. Jamal put 600 pennies into 6 equal rolls. How many pennies were in each roll?

_____

24. Sela has 6 times as many coins now as she had 4 months ago. If Sela has 240 coins now, how many did she have 4 months ago?

_____

SHOW YOUR WORK

25. **H.O.T.** Chip collected 2,090 dimes. Sue collected 1,910 dimes. They divided all their dimes into 8 equal stacks. How many dimes are in each stack?

_____

_____

26. **Write Math** ▶ Mr. Roberts sees a rare 1937 penny. The cost of the penny is $210. If he saves $3 a week, will Mr. Roberts have enough money to buy the penny in one year? **Explain**.

_____

_____

_____

27. Carine sold $320 worth of cookies. If each box of cookies costs $4, how many boxes did she sell?

_____

28. **Test Prep** Which number sentence is not true?

Ⓐ 150 ÷ 5 = 30   Ⓒ 4,500 ÷ 9 = 500

Ⓑ 400 ÷ 8 = 500   Ⓓ 5,600 ÷ 7 = 800

Connect to Science

Insect Flight

True flight is shared only by insects, bats, and birds. Flight in insects varies from the clumsy flight of some beetles to the acrobatic moves of dragonflies.

The wings of insects are not moved by muscles attached to the wings. Muscles in the middle part of the body, or thorax, move the wings. The thorax changes shape as the wings move.

| Insect Wing Beats in 3 Minutes | |
Insect	Approximate Number of Wing Beats
Aeschnid Dragonfly	6,900
Damselfly	2,700
Large White Butterfly	2,100
Scorpion Fly	5,000

29. About how many times does a damselfly's wings beat in 1 minute?

_____

30. About how many times do a scorpion fly's wings beat in 6 minutes?

_____

31. **H.O.T.** In one minute, about how many more times do a damselfly's wings beat than a large white butterfly's wings?

_____

32. **What's the Question?** The answer is about 2,300 times.

_____

_____

Name _____

Estimate Quotients Using Compatible Numbers

Essential Question How can you use compatible numbers to estimate quotients?

🔑 UNLOCK the Problem REAL WORLD

A horse's heart beats 132 times in 3 minutes. About how many times does it beat in 1 minute?

You can use compatible numbers to estimate quotients.

Compatible numbers are numbers that are easy to compute mentally.

- Will a horse's heart beat more or fewer than 132 times in 1 minute?

_____

- What operation will you use to solve the problem?

_____

🔑 Example 1 Estimate. 132 ÷ 3

STEP 1 Find a number close to 132 that divides easily by 3. Use basic facts.

12 ÷ 3 is a basic fact. 120 divides easily by 3.

15 ÷ 3 is a basic fact. 150 divides easily by 3.

Think: Choose 120 because it is closer to 132.

STEP 2 Use place value.

120 = _____ tens

12 ÷ 3 = _____

12 tens ÷ 3 = _____ tens

120 ÷ 3 = _____

So, a horse's heart beats about _____ times a minute.

🔑 Example 2 Use compatible numbers to find two estimates that the quotient is between. 1,382 ÷ 5

STEP 1 Find two numbers close to 1,382 that divide easily by 5.

_____ ÷ 5 is a basic fact. 1,000 divides easily by 5.

_____ ÷ 5 is a basic fact. 1,500 divides easily by 5.

1,382 is between _____ and _____.

STEP 2 Divide each number by 5. Use place value.

1,000 ÷ 5

_____ hundreds ÷ 5 = _____ hundreds, or _____

1,500 ÷ 5

_____ hundreds ÷ 5 = _____ hundreds, or _____

So, 1,382 ÷ 5 is between _____ and _____.

Math Talk **MATHEMATICAL PRACTICES** **Explain** which estimate you think is more reasonable.

Share and Show

1. Estimate. 1,718 ÷ 4

Think: What number close to 1,718 is easy to divide by 4?

_____ is close to 1,718.

What basic fact can you use? _____ ÷ 4

_____ is close to 1,718.

What basic fact can you use? _____ ÷ 4

Choose 1,600 because _____.

16 ÷ 4 = _____

1,600 ÷ _____ = _____

1,718 ÷ 4 is about _____

> **MATHEMATICAL PRACTICES**
>
> **Math Talk** Explain how your estimate might change if the problem were 1,918 ÷ 4.

Use compatible numbers to estimate the quotient.

2. 455 ÷ 9

3. 1,509 ÷ 3

4. 176 ÷ 8

5. 2,795 ÷ 7

On Your Own

Use compatible numbers to estimate the quotient.

6. 163 ÷ 2

7. 500 ÷ 7

8. 1,421 ÷ 5

9. 2,642 ÷ 8

Use compatible numbers to find two estimates that the quotient is between.

10. 5,321 ÷ 6

11. 1,765 ÷ 6

12. 1,189 ÷ 3

13. 2,110 ÷ 4

H.O.T. **Algebra** Estimate to compare. Write <, >, or =.

14. 613 ÷ 3 ◯ 581 ÷ 2

15. 364 ÷ 4 ◯ 117 ÷ 6

16. 2,718 ÷ 8 ◯ 963 ÷ 2

_____ _____
estimate estimate

_____ _____
estimate estimate

_____ _____
estimate estimate

Problem Solving REAL WORLD

Use the table for 17–20.

17. About how many times does a chicken's heart beat in 1 minute?

_____

18. **H.O.T.** About how many times does a cow's heart beat in 2 minutes?

_____

19. **H.O.T.** About how many times faster does a cow's heart beat than a whale's?

_____

20. **Write Math** ▶ **What's the Question?** The answer is about 100 beats in 1 minute.

_____

_____

_____

21. Jamie and his two brothers divided a package of 125 toy cars equally. About how many cars did each of them receive?

_____

22. **Test Prep** A monkey's heart beats 1,152 times in 6 minutes. Which is the best estimate of the number of times its heart beats in 1 minute?

Ⓐ 100

Ⓑ 200

Ⓒ 1,000

Ⓓ 2,000

Animal Heartbeats in 5 Minutes

Animal	Number of Heartbeats
Whale	31
Cow	325
Pig	430
Dog	520
Chicken	1,375

SHOW YOUR WORK

Cause and Effect

The reading skill *cause and effect* can help you understand how one detail in a problem is related to another detail.

Bike Shop Layaway Plans

Plan A	3 months (3 equal payments)
Plan B	6 months (6 equal payments)

Chet wants to buy a new bike that costs $276. Chet mows his neighbor's lawn for $15 each week. Since Chet does not have money saved, he needs to decide which layaway plan he can afford to buy the new bike.

Cause: Chet does not have money saved to purchase the bike.		**Effect:** Chet will have to decide which layaway plan he can afford to purchase the bike.

Which plan should Chet choose?

3-month layaway:	6-month layaway:
$276 ÷ 3	$276 ÷ 6
Estimate.	Estimate.
$270 ÷ 3 _____	$300 ÷ 6 _____

Chet earns $15 each week. Since there are usually 4 weeks in a month, multiply to see which payment he can afford.

$$\$15 \times 4 = \underline{\hspace{2cm}}$$

So, Chet can afford the _____ layaway plan.

Use estimation to solve.

23. Sofia wants to buy a new bike that costs $214. Sofia helps her grandmother with chores each week for $18. Estimate to find which layaway plan Sofia should choose and why.

_____

_____

_____

24. **Write Math** ▶ Describe a situation when you have used cause and effect to help you solve a math problem.

_____

_____

_____

_____

Name _____

Division and the Distributive Property

Essential Question How can you use the Distributive Property to find quotients?

Investigate

Materials ■ color pencils ■ grid paper

You can use the Distributive Property to break apart numbers to make them easier to divide.

The Distributive Property of division says that dividing a sum by a number is the same as dividing each addend by the number and then adding the quotients.

A. Outline a rectangle on a grid to model 69 ÷ 3.

Shade columns of 3 until you have 69 squares.

How many groups of 3 can you make? _____

B. Think of 69 as 60 + 9. Break apart the model into two rectangles to show (60 + 9) ÷ 3. Label and shade the smaller rectangles. Use two different colors.

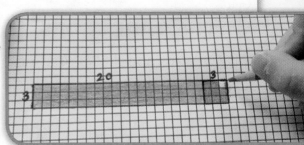

C. Each rectangle models a division.

69 ÷ 3 = (_____ ÷ 3) + (_____ ÷ 3)

= _____ + _____

= _____

D. Outline another model to show 68 ÷ 4.

How many groups of 4 can you make? _____

E. Think of 68 as 40 + 28. Break apart the model, label, and shade to show two divisions.

68 ÷ 4 = (_____ ÷ 4) + (_____ ÷ 4)

= _____ + _____

= _____

Draw Conclusions

1. **Explain** how each small rectangle models a quotient and a product in Step C.

   _____

   _____

   _____

2. **Compare** your answer in Step A to the final quotient in Step C. What can you conclude?

   _____

   _____

3. **H.O.T.** **Evaluate** To find the quotient $91 \div 7$, would you break up the dividend into $90 + 1$ or $70 + 21$? Explain.

   _____

   _____

Make Connections

Math Talk MATHEMATICAL PRACTICES Describe another way you could use the Distributive Property to solve $68 \div 4$.

You can also model $68 \div 4$ using base-ten blocks.

STEP 1 Model 68.

$68 =$ _____ $+$ _____

STEP 2 Divide the longs into 4 equal groups. 4 longs divide into 4 equal groups with 2 longs left. Regroup 2 longs as 20 small cubes. Divide them evenly among the 4 groups.

$60 \div 4 =$ _____

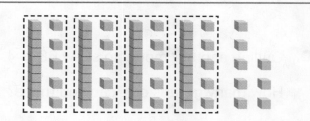

STEP 3 Divide the 8 small cubes into the 4 equal groups.

$8 \div 4 =$ _____

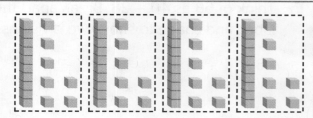

So, $68 \div 4 = (60 \div 4) + (8 \div 4) =$ _____ $+$ _____ $=$ _____

Name _____

Model the division on the grid.

 **1.** $26 \div 2 = ($ _____ $\div\ 2) + ($ _____ $\div\ 2)$

$= $ _____ $+$ _____

$= $ _____

2. $45 \div 3 = ($ _____ $\div\ 3) + ($ _____ $\div\ 3)$

$= $ _____ $+$ _____

$= $ _____

Find the quotient.

 **3.** $86 \div 2$

$= ($ _____ $\div\ 2) + ($ _____ $\div\ 2)$

$= $ _____ $+$ _____

$= $ _____

4. $208 \div 4$

$= ($ _____ $\div\ 4) + ($ _____ $\div\ 4)$

$= $ _____ $+$ _____

$= $ _____

Use base-ten blocks to model the quotient.
Then record the quotient.

5. $88 \div 4 = $ _____

6. $36 \div 3 = $ _____

7. $186 \div 6 = $ _____

8. $96 \div 8 = $ _____

9. $189 \div 9 = $ _____

10. $54 \div 2 = $ _____

11. $707 \div 7 = $ _____

12. $255 \div 5 = $ _____

13. $612 \div 6 = $ _____

14. **Write Math** ▶ **Explain** how you can model finding quotients using the Distributive Property.

_____

_____

_____

_____

Problem Solving  REAL WORLD

H.O.T. Pose a Problem

15. Christelle went to a gift shop. The shop sells candles in a variety of sizes and colors. The picture shows a display of candles.

Write a problem that can be solved using the picture.

Pose a problem.

_____

_____

_____

_____

_____

_____

_____

_____

Solve your problem.

• **Describe** how you could change the problem by changing the number of rows of candles. Then solve the problem.

_____

_____

_____

_____

✓ Mid-Chapter Checkpoint

▶ **Vocabulary**

Choose the best term from the box to complete the sentence.

Vocabulary

counting numbers

compatible numbers

multiple

remainder

1. A number that is the product of a number and a counting

   number is called a _____. (p. 137)

2. Numbers that are easy to compute mentally are called

   _____. (p. 153)

3. When a number cannot be divided evenly, the amount

   left over is called the _____. (p. 142)

▶ **Concepts and Skills**

Divide. Draw a quick picture to help.

4. $26 \div 3$ _____

5. $19 \div 4$ _____

Use basic facts and place value to find the quotient.

6. $810 \div 9 =$ _____

7. $210 \div 7 =$ _____

8. $3,000 \div 6 =$ _____

Use compatible numbers to estimate the quotient.

9. $635 \div 9$

   _____

10. $412 \div 5$

    _____

11. $490 \div 8$

    _____

Use grid paper or base-ten blocks to model the quotient. Then record the quotient.

12. $63 \div 3 =$ _____

13. $85 \div 5 =$ _____

14. $168 \div 8 =$ _____

Fill in the bubble completely to show your answer.

15. Ana has 296 coins in her coin collection. She put the same number of coins in each of 7 jars. About how many coins are in each jar?

    (A) about 20 coins

    (B) about 40 coins

    (C) about 200 coins

    (D) about 400 coins

16. Which two estimates is the quotient $345 \div 8$ between?

    (A) 40 and 50   (C) 400 and 500

    (B) 50 and 60   (D) 500 and 600

17. A peanut vendor had 640 bags of peanuts. She sold the same number of bags of peanuts at each of 8 baseball games. How many bags of peanuts did she sell at each game?

    (A) 8   (C) 80

    (B) 10   (D) 800

18. There are 4 students on a team for a relay race. How many teams can be made from 27 students?

    (A) 3   (C) 6

    (B) 4   (D) 7

19. Eight teams of high school students helped clean up trash in the community. Afterwards, they shared 23 pizzas equally. How many pizzas did each team get?

    (A) 2   (C) $2\frac{7}{8}$

    (B) $2\frac{3}{8}$ (D) 3

Name _____

Divide Using Repeated Subtraction

Essential Question How can you use repeated subtraction and multiples to find quotients?

Investigate

Materials ■ counters ■ grid paper

John is building a backyard pizza oven with an arch opening. He has 72 bricks. He will place 6 bricks at a time as he builds the oven. If he arranges the bricks in piles of 6, how many piles will he have?

You can use repeated subtraction to divide $72 \div 6$.

A. Begin with 72 counters. Subtract 6 counters.

How many are left? _____

B. Record the subtraction on grid paper as shown. Record the number of counters left and the number of times you subtracted.

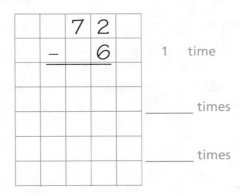

	7	2
−		6

    1   time

    _____ times

    _____ times

C. Can you reach zero evenly? Explain.

_____

_____

D. Count the number of times you subtracted 6 counters. _____

So, there are _____ piles of 6 bricks.

Draw Conclusions

1. **Explain** the relationship between the divisor, the dividend, the quotient, and the number of times you subtracted the divisor from the dividend.

   _____

   _____

2. **Synthesize** What happens if you subtract multiples of 6? Complete the example at the right.

   $$6 \overline{)72}$$
   $$-60 \leftarrow \boxed{} \times 6 \quad 10$$
   $$-12 \leftarrow \boxed{} \times 6 + \underline{}$$

   - What multiples of 6 did you use? How did you use them?

     _____

     _____

   - What numbers did you add? Why?

     _____

     _____

   - How did using multiples of the divisor help you?

     _____

     _____

3. **H.O.T. Justify** Why should you subtract 10×6 and not 9×6 or 20×6?

   _____

   _____

   _____

Math Talk   MATHEMATICAL PRACTICES
Explain how subtracting counters and counting back on a number line help you divide.

Make Connections

Another way to divide by repeated subtraction is to use a number line. Count back by 4s from 52 to find $52 \div 4$.

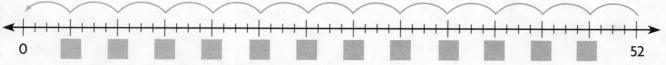

0   52

How many equal groups of 4 did you subtract? _____

So, $52 \div 4 =$ _____

Name _____

Share and Show

Use repeated subtraction to divide.

1. 84 ÷ 7 _____

2. 60 ÷ 4 _____

3. 91 ÷ 8 _____

Draw a number line to divide.

4. 65 ÷ 5 = _____

5. 78 ÷ 6 = _____

6. 91 ÷ 7 = _____

7. **Write Math** Can you divide 32 by 3 evenly? Use the number line to explain your answer.

_____

_____

_____

🔑 UNLOCK the Problem REAL WORLD

8. A new playground will be 108 feet long. Builders need to allow 9 feet of space for each piece of climbing equipment. They want to put as many climbers along the length of the playground as possible. How many climbers can they place?

a. What are you asked to find?

_____

_____

b. How can you use repeated subtraction to solve the problem?

_____

_____

c. Tell why you might use multiples of the divisor to solve the problem.

_____

_____

d. Show steps to solve the problem.

e. Complete the sentences.

There are _____ equal parts of the

playground, each _____ feet long.

So, _____ climbers can fit along the length of the playground.

---

9. There are 128 students in the fourth grade. Half of the students can use the playground at the same time. How many students is that?

_____

10. Test Prep An architect designed the school auditorium. There are 84 seats in Section A. Each row has 6 seats. How many rows of seats are in Section A?

Ⓐ 4 Ⓑ 14 Ⓒ 24 Ⓓ 60

Divide Using Partial Quotients

Essential Question How can you use partial quotients to divide by 1-digit divisors?

 **UNLOCK the Problem** REAL WORLD

At camp, there are 5 players on each lacrosse team. If there are 125 people on lacrosse teams, how many teams are there?

- Underline what you are asked to find.
- Circle what you need to use.
- What operation can you use to find the number of teams?

_____

One Way Use partial quotients.

In the **partial quotient** method of dividing, multiples of the divisor are subtracted from the dividend and then the partial quotients are added together.

Divide. 125 ÷ 5   **Write.** 5)125

STEP 1

Start by subtracting a greater multiple, such as 10 times the divisor. For example, you know that you can make at least 10 teams of 5 players.

Continue subtracting until the remaining number is less than the multiple, 50.

STEP 2

Subtract smaller multiples, such as 5, 2, or 1 times the divisor until the remaining number is less than the divisor. In other words, keep going until you no longer have enough players to make a team.

Then add the partial quotients to find the quotient.

So, there are _____ lacrosse teams.

Partial Quotients

5)125   ↓

− ▭   10 × ____   10

− ▭   10 × ____   10

− ▭   5 × ____   + 5

Math Talk **Explain** how you found the total number of teams after finding the partial quotients.

🔑 Another Way Use rectangular models to record the partial quotients.

Jarod and Mi also found the number of teams using partial quotients. They recorded the partial quotients using rectangular models. They each still had 25 as the quotient.

Jarod

| 5 | 125 |

10	
5	50	75

$$\begin{array}{r} 125 \\ - \quad \\ \hline 75 \end{array}$$

10	10	
5	50	50	25

$$\begin{array}{r} 75 \\ - \quad \\ \hline 25 \end{array}$$

10	10	5
5	50	50	25

$$\begin{array}{r} 25 \\ - \quad \\ \hline 0 \end{array}$$

10 + 10 + 5 = _____

Mi

| 5 | 125 |

20	
5	100	25

$$\begin{array}{r} 125 \\ - \quad \\ \hline 25 \end{array}$$

20	5
5	100	25

$$\begin{array}{r} 25 \\ - \quad \\ \hline 0 \end{array}$$

20 + 5 = _____

MATHEMATICAL PRACTICES

Math Talk Explain why you might prefer to use one method rather than the other.

Share and Show 🖊️ MATH BOARD ·

1. Lacrosse is played on a field 330 ft long. How many yards long is a lacrosse field? (3 feet = 1 yard)

Divide. Use partial quotients.

$$3\overline{)330}$$

$-$ ___ 100 × ☐ 100

$-$ ___ 10 × ☐ + 10

So, the lacrosse field is _____ yards long.

Name _____

Divide. Use partial quotients.

✓ **2.** 3)225

Divide. Use rectangular models to record the partial quotients.

✓ **3.** 428 ÷ 4 = _____

Math Talk MATHEMATICAL PRACTICES

Explain how you could solve Problems 2 and 3 a different way.

On Your Own ·····································

Divide. Use partial quotients.

4. 9)198

5. 7)259

6. 8)864

7. 6)738

Divide. Use rectangular models to record the partial quotients.

8. 328 ÷ 2 = _____

9. 475 ÷ 5 = _____

10. 219 ÷ 3 = _____

11. 488 ÷ 4 = _____

Practice: Copy and Solve Divide. Use either way to record the partial quotients.

12. 875 ÷ 5

13. 372 ÷ 2

14. 252 ÷ 6

15. 429 ÷ 3

16. 568 ÷ 8

17. 504 ÷ 7

18. 624 ÷ 4

19. 819 ÷ 9

© Houghton Mifflin Harcourt Publishing Company

Problem Solving · REAL WORLD

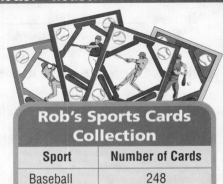

Use the table for 20–22.

20. Rob wants to put 8 baseball cards on each page in an album. How many pages will he fill?

_____

21. Rob filled 9 plastic boxes with basketball cards with the same number of cards in each box. How many cards did he put in each box?

_____

22. **H.O.T.** Rob filled 3 fewer plastic boxes with football cards than basketball cards. How many boxes did he fill? How many football cards were in each box?

_____

Rob's Sports Cards Collection

Sport	Number of Cards
Baseball	248
Basketball	189
Football	96
Hockey	64

23. A professional game of lacrosse has 60 minutes of playing time. It is split into 4 equal periods of time. How many minutes are in each period?

_____

24. **H.O.T.** **Write Math** ▶ What is the least number you can divide by 5 to get a three-digit quotient? Explain how you found your answer.

_____

_____

_____

_____

_____

· · · · · · **SHOW YOUR WORK** · · · · · ·

25. **Test Prep** There are 126 students who signed up to learn how to play lacrosse. If there are 6 students in each group, how many groups are there?

Ⓐ 12   Ⓒ 21

Ⓑ 20   Ⓓ 120

FOR MORE PRACTICE:
Standards Practice Book, pp. P83–P84

Model Division with Regrouping

Essential Question How can you use base-ten blocks to model division with regrouping?

Investigate

Materials ▪ base-ten blocks

The librarian wants to share 54 books equally among 3 classes. How many books will she give to each class?

A. Draw 3 circles to represent the classes. Then use base-ten blocks to model 54. Show 54 as 5 tens and 4 ones.

B. Share the tens equally among the 3 groups.

C. If there are any tens left, regroup them as ones. Share the ones equally among the 3 groups.

D. There are _____ ten(s) and _____ one(s) in each group.

So, the librarian will give _____ books to each class.

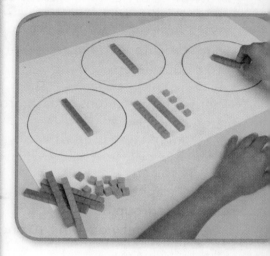

Draw Conclusions ....

1. **H.O.T.** **Explain** why you needed to regroup in Step C.

_____

_____

2. **Apply** How you can use base-ten blocks to find the quotient of $92 \div 4$?

_____

_____

_____

_____

Make Connections . . . .

Use the quick picture at the bottom of the page to help you divide.
Record each step.

Find 76 ÷ 3.

STEP 1
Model 76 as 7 tens 6 ones.
Draw three circles to represent equal groups.

STEP 2
Share the 7 tens equally among the 3 groups.
Cross out the tens you use.

There are _____ tens in each group.

_____ tens were used. There is _____ ten left over.

tens in each group

tens used

ten left over

STEP 3
One ten cannot be shared among 3 groups
without regrouping.
Regroup 1 ten by drawing 10 ones.

There are now _____ ones to share.

$$\begin{array}{r} 2 \\ 3\overline{)76} \\ -6 \end{array}$$

ones to share

STEP 4
Share the ones equally among the 3 groups.
Cross out the ones you use.

There are _____ ones in each group.

_____ ones were used. There is _____ one left over.

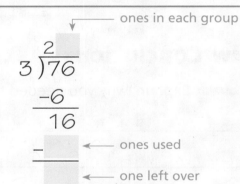

ones in each group

ones used

one left over

There are 3 groups of _____ and _____ left over.

So, for 76 ÷ 3, the quotient is _____ and the remainder is _____.

This can be written as _____.

MATHEMATICAL PRACTICES

Math Talk Why do you share tens equally among groups before sharing ones?

Name _____

Share and Show

Divide. Use base-ten blocks.

1. $48 \div 3$ _____

2. $84 \div 4$ _____

✓ **3.** $72 \div 5$ _____

Divide. Draw quick pictures. Record the steps.

4. $59 \div 2$ _____

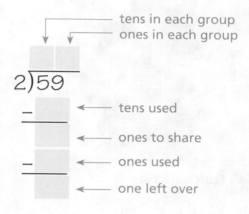

✓ **5.** $84 \div 3$ _____

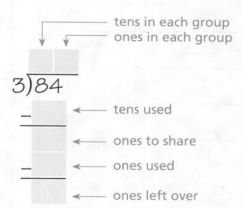

6. **Write Math** ▸ **Explain** why you did not need to regroup in Exercise 2.

_____

_____

Problem Solving REAL WORLD

Sense or Nonsense?

7. Angela and Zach drew quick pictures to find 68 ÷ 4. Whose quick picture makes sense? Whose quick picture is nonsense? **Explain** your reasoning.

I drew 1 ten and 2 ones in each group.

I drew 1 ten and 7 ones in each group.

Angela's Quick Picture	**Zach's Quick Picture**

_____

_____

_____

_____

_____

_____

_____

_____

_____

_____

8. Analyze What did Angela forget to do after she shared the tens equally among the 4 groups?

_____

_____

Place the First Digit

Essential Question How can you use place value to know where to place the first digit in the quotient?

 **UNLOCK the Problem**  REAL WORLD

Jaime took 144 photos on a digital camera.
The photos are to be placed equally in 6 photo albums.
How many photos will be in each album?

- Underline what you are asked to find.
- Circle what you need to use.

🔒 **Example 1** Divide. 144 ÷ 6

STEP 1 Use place value to place the first digit.
Look at the hundreds in 144.
1 hundred cannot be shared among 6 groups without regrouping.
Regroup 1 hundred as 10 tens.

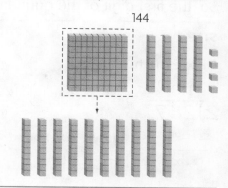
144

Now there are _____ tens to share among 6 groups.

The first digit of the quotient will be in the _____ place.

STEP 2 Divide the tens.

$$\begin{array}{r} 2 \\ 6\overline{)144} \\ - \end{array}$$

Divide. 14 tens ÷ 6

Multiply. 6 × 2 tens

Subtract. 14 tens − 12 tens
Check. 2 tens cannot be shared among 6 groups without regrouping.

STEP 3 Divide the ones.
Regroup 2 tens as 20 ones.

Now there are _____ ones to share among 6 groups.

$$\begin{array}{r} 24 \\ 6\overline{)144} \\ -12\downarrow \\ \hline 24 \\ - \end{array}$$

Divide. _____ ones ÷ _____

Multiply. _____ × _____ ones

Subtract. _____ ones − _____ ones
Check. 0 ones cannot be shared among 6 groups.

Math Idea
After you divide each place, the remainder should be less than the divisor.

Math Talk MATHEMATICAL PRACTICES
Explain how the answer would change if Jaime had 146 photos.

So, there will be _____ photos in each album.

🔑 Example 2 Divide. 287 ÷ 2

Reggie has 287 photographs of animals. If he wants to put the photos into 2 groups of the same size, how many photos will be in each group?

STEP 1

Use place value to place the first digit.
Look at the hundreds in 287.
2 hundreds can be shared between 2 groups.

So, the first digit of the quotient will be in the _____ place.

STEP 2

Divide the hundreds.

$$\begin{array}{r} 1 \\ 2\overline{)287} \\ - \\ \hline \end{array}$$

Divide. 2 hundreds ÷ 2

Multiply. 2 × 1 hundred

Subtract. 2 hundreds − 2 hundreds.

0 hundreds are left.

STEP 3

Divide the tens.

$$\begin{array}{r} 14 \\ 2\overline{)287} \\ -2\downarrow \\ \hline 0 \\ - \\ \hline \end{array}$$

Divide. _____ tens ÷ _____

Multiply. _____ × _____ tens

Subtract. _____ tens − _____ tens 0 tens are left.

STEP 4

Divide the ones.

$$\begin{array}{r} 143\,r1 \\ 2\overline{)287} \\ -2 \\ \hline 08 \\ -8\downarrow \\ \hline 07 \\ - \\ \hline \end{array}$$

Divide. _____ ones ÷ _____

Multiply. _____ × _____ ones

Subtract. _____ ones − _____ ones

1 one cannot be equally shared between 2 groups.

So, there will be _____ photos in each group with 1 photo left.

Name _____

Share and Show

1. There are 452 pictures of dogs in 4 equal groups. How many pictures are in each group? **Explain** how you can use place value to place the first digit in the quotient.

$$4\overline{)452}$$

_____

_____

Divide.

✓ **2.** $4\overline{)166}$ ✓ **3.** $5\overline{)775}$

Math Talk MATHEMATICAL PRACTICES
Explain how you placed the first digit of the quotient in Exercise 2.

On Your Own

Divide.

4. $4\overline{)284}$ **5.** $5\overline{)394}$ **6.** $3\overline{)465}$ **7.** $8\overline{)272}$

8. $2\overline{)988}$ **9.** $3\overline{)504}$ **10.** $6\overline{)734}$ **11.** $4\overline{)399}$

Practice: Copy and Solve Divide.

12. $516 \div 2$ **13.** $516 \div 3$ **14.** $516 \div 4$ **15.** $516 \div 5$

16. H.O.T. Look back at your answers to Exercises 12–15. What happens to the quotient when the divisor increases? **Explain.**

_____

_____

© Houghton Mifflin Harcourt Publishing Company

UNLOCK the Problem REAL WORLD

17. Nan wants to put 234 pictures in an album with a blue cover. How many full pages will she have in her album?

a. What do you need to find?

_____

b. How will you use division to find the number of full pages?

_____

_____

_____

Photo Albums

Color of cover	Pictures per page
Blue	4
Green	6
Red	8

c. Show the steps you will use to solve the problem.

d. Complete the following sentences.

Nan has _____ pictures.

She wants to put the pictures in an album

with pages that each hold _____ pictures.

She will have an album with _____ full

pages and _____ pictures on another page.

18. Juan wants to put his 672 pictures in an album with a green cover. How many full pages will he have in his album?

19. Test Prep Kat wants to put her 485 pictures in an album with a red cover. She uses division to find out how many full pages she will have. In which place is the first digit of the quotient?

Ⓐ thousands

Ⓑ hundreds

Ⓒ tens

Ⓓ ones

Divide by 1-Digit Numbers

Essential Question How can you divide multidigit numbers and check your answers?

UNLOCK the Problem REAL WORLD

Students in the third, fourth, and fifth grades made 525 origami animals to display in the library. Each grade made the same number of animals. How many animals did each grade make?

🔑 Example 1 Divide. 525 ÷ 3

STEP 1 Use place value to place the first digit.
Look at the hundreds in 525.
5 hundreds can be shared among
3 groups without regrouping.
The first digit of the quotient will be in the _____ place.

STEP 2 Divide the hundreds.

$$3\overline{)525}^{1}$$
$$-$$

Divide. Share _____ hundreds equally among _____ groups.

Multiply. _____ × _____

Subtract. _____ − _____.

Check. _____ hundreds cannot be shared among 3 groups without regrouping.

Math Talk MATHEMATICAL PRACTICES
At the checking step, what would you do if the number is greater than the divisor?

STEP 3 Divide the tens.

$$3\overline{)525}^{17}$$
$$\underline{-3}\downarrow$$
$$22$$
$$-$$

Divide. Share _____ equally among _____ groups.

Multiply. _____

Subtract. _____ − _____

Check. _____
_____
_____.

STEP 4 Divide the ones.

$$3\overline{)525}^{175}$$
$$\underline{-3}\downarrow$$
$$22$$
$$\underline{-21}\downarrow$$
$$15$$
$$-$$

Divide. Share _____ equally among _____ groups.

Multiply. _____

Subtract. _____ − _____

Check. _____ are left.

So, each class made _____ origami animals.

There are 8,523 sheets of origami paper to be divided equally among 8 schools. How many sheets of origami paper will each school get?

 Example 2 Divide. 8,523 ÷ 8

STEP 1 Use place value to place the first digit.

Look at the thousands in 8,523.
8 thousands can be shared among
8 groups without regrouping.

The first digit of the quotient will be

in the _____ place.

STEP 2 Divide the thousands. _____

STEP 3 Divide the hundreds. _____

STEP 4 Divide the tens. _____

STEP 5 Divide the ones.

So, each school will get _____ sheets of origami paper.

There will be _____ sheets left.

$$8\overline{)8,523}$$

 ERROR Alert

Place a zero in the quotient when a place in the dividend cannot be divided by the divisor.

CONNECT Division and multiplication are inverse operations. You can use multiplication to check your answer to a division problem.

Multiply the quotient by the divisor. If there is a remainder, add it to the product. The result should equal the dividend.

Divide.

quotient → 1,065 r3 ← remainder
divisor → 8)8,523 ← dividend

Check.

```
      1,065    ← quotient
    ×     8    ← divisor
      8,520
    +     3    ← remainder
      8,523    ← dividend
```

The check shows that the division is correct.

Name _____

Share and Show

1. Ollie used 852 beads to make 4 bracelets. He put the same number of beads on each bracelet. How many beads does each bracelet have? Check your answer.

Divide.

Check.

So, each bracelet has _____ beads.

MATHEMATICAL PRACTICES

Math Talk Explain how you could check if your quotient is correct.

Divide and check.

2. $2\overline{)394}$

✅ 3. $2\overline{)803}$

✅ 4. $4\overline{)3,448}$

On Your Own

Divide and check.

5. $2\overline{)816}$

6. $4\overline{)709}$

7. $3\overline{)267}$

8. $6\overline{)1,302}$

9. $8\overline{)9,232}$

10. $9\overline{)1,020}$

H.O.T.   Algebra Find the unknown number.

11. $n \div 3 = 315$

$n =$ _____

12. $n \div 4 = 1,225$

$n =$ _____

13. $185 = n \div 5$

$n =$ _____

Problem Solving REAL WORLD

Use the table for 14–16.

14. Four teachers bought 10 origami books and 100 packs of origami paper for their classrooms. They will share the cost of the items equally. How much should each teacher pay?

_____

15. **H.O.T.**   **Write Math** Six students shared equally the cost of 18 of one of the items in the chart. Each student paid $24. What item did they buy? **Explain** how you found your answer.

_____

_____

_____

_____

16. Ms. Alvarez has $1,482 to spend on origami paper. How many packs can she buy?

_____

17. Evan made origami cranes with red, blue, and yellow paper. The number of cranes in each color is the same. If there are 342 cranes, how many of them are blue or yellow?

_____

18. **Test Prep** An artist made 515 origami animals in 5 days. She made the same number of animals each day. How many origami animals did she make each day?

(A) 13

(B) 103

(C) 510

(D) 2,060

The Craft Store

Item	Price
Origami Book	$24 each
Origami Paper	$6 per pack
Origami Kit	$8 each

· · · · · **SHOW YOUR WORK** · · · · ·

Name _____

Problem Solving • Multistep Division Problems

Essential Question How can you use the strategy *draw a diagram* to solve multistep division problems?

🔑 UNLOCK the Problem REAL WORLD

Lucia and her dad will prepare corn for a community picnic. There are 3 bags of corn. Each bag holds 32 ears of corn. When the corn is cooked, they want to divide the corn equally among 8 serving plates. How many ears of corn should they put on each of 8 serving plates?

Read the Problem	Solve the Problem
What do I need to find? I need to find the number of _____ that will go on each plate. **What information do I need to use?** _____ bags with _____ ears in each bag. The total ears are divided equally into _____ groups. **How will I use the information?** I will make a bar model for each step to visualize the information. Then I will _____ to find the total and _____ to find the number for each plate.	I can draw bar models to visualize the information given and then decide how to find how many ears of corn should go on a plate. First, I will model and multiply to find the total number of ears of corn. <table><tr><td>32</td><td>32</td><td>32</td></tr></table> Then I will model and divide to find how many ears of corn should go on each plate. <table><tr><td></td><td></td><td></td><td></td><td></td><td></td><td></td><td></td></tr></table> 96

1. How many ears of corn should go on each plate? _____

2. How can you check your answer? _____

_____

🔑 Try Another Problem

There are 8 dinner rolls in a package. How many packages will be needed to feed 64 people if each person has 2 dinner rolls?

Read the Problem	Solve the Problem
What do I need to find?	
What information do I need to use?	
How will I use the information?	

3. How many packages of rolls will be needed? _____

4. How did drawing a bar model help you solve the problem?

_____

_____

_____

© Houghton Mifflin Harcourt Publishing Company

Math Talk MATHEMATICAL PRACTICES
Describe another method you could have used to solve the problem.

Name _____

Share and Show MATH BOARD

🔑 **UNLOCK the Problem** Tips
✓ Use the Problem Solving MathBoard.
✓ Underline important facts.
✓ Choose a strategy you know.

1. A firehouse pantry has 52 cans of vegetables and 74 cans of soup. Each shelf holds 9 cans. What is the least number of shelves needed for all the cans?

   First, draw a bar model for the total number of cans.

   Next, add to find the total number of cans.

   Then, draw a bar model to show the number of shelves needed.

   Finally, divide to find the number of shelves needed.

Math Talk MATHEMATICAL PRACTICES
Explain how you could check that your answer is correct.

••• **SHOW YOUR WORK** •••

So, _____ shelves are needed to hold all of the cans.

2. **H.O.T.** **What if** 18 cans fit on a shelf? What is the least number of shelves needed? **Describe** how your answer would be different.

   _____

   _____

3. Julio's dad bought 10 dozen potatoes. The potatoes were equally divided into 6 bags. How many potatoes are in each bag?

   _____

4. Ms. Johnson is in charge of decorations for a party. She bought 6 bags of balloons. Each bag has 25 balloons. She fills all the balloons and puts 5 balloons in each bunch. How many bunches can she make?

   _____

On Your Own..............

Choose a STRATEGY

Act It Out
Draw a Diagram
Find a Pattern
Make a Table or List
Solve a Simpler Problem

5. At the garden shop, each small tree costs $125 and each large tree costs $225. How much will 3 small trees and 1 large tree cost?

_____

6. **H.O.T.** An adult's dinner costs $8. A family of 2 adults and 2 children pays $26 for their dinners. How much does a child's dinner cost? **Explain.**

_____

_____

Use the table for 7–8.

7. **Write Math** ▶ Maria bought 80 ounces of apples. She needs 10 apples to make a pie. How many apples will be left over? **Explain.**

_____

_____

_____

Fruit	Average weight
Peach	6 ounces
Apple	5 ounces
Plum	2 ounces

8. Molly put 4 pieces of fruit in a bag. The bag weighs 19 ounces. How many of each kind of fruit are in the bag?

_____

9. The garden warehouse delivered 1,500 pounds of topsoil in 5-pound bags to the garden shop. The garden shop sold half of the bags the same day they were delivered. How many bags does the garden shop have left to sell?

_____

10. **Test Prep** Ben collected 43 cans and some bottles. He received 5¢ for each can or bottle. If Ben received a total of $4.95, how many bottles did he collect?

   (A) 56   (B) 99   (C) 560   (D) 990

FOR MORE PRACTICE:
Standards Practice Book, pp. P91–P92

Name _____

Chapter Review/Test

▶ **Vocabulary**

Choose the best term from the box.

Vocabulary
compatible numbers
partial quotient
remainder

1. When a number cannot be divided evenly, the amount

   left over is called the _____. (p. 142)

2. You use the _____ method of dividing when
   multiples of the divisor are subtracted from the dividend and
   then the quotients are added together. (p. 167)

▶ **Concepts and Skills**

**Use grid paper or base-ten blocks to model the quotient.
Then record the quotient.**

3. 96 ÷ 6 = _____   4. 86 ÷ 2 = _____   5. 155 ÷ 5 = _____

**Find two numbers the quotient is between.
Then estimate the quotient.**

6. 787 ÷ 2   7. 391 ÷ 6   8. 789 ÷ 8

   _____   _____   _____

   _____   _____   _____

Divide.

9. 3)987   10. 7)501   11. 5)153

12. 4)808   13. 6)8,348   14. 8)4,897

GO Online Assessment Options
Chapter Test

Fill in the bubble completely to show your answer.

15. There are 96 tourists who have signed up to tour the island. The tourists are assigned to 6 equal-size groups. How many tourists are in each group?

    (A) 1 r3

    (B) 1 r6

    (C) 11

    (D) 16

16. Maria needs to share the base-ten blocks equally among 4 equal groups.

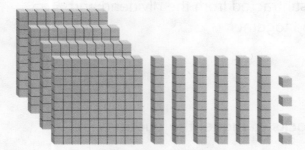

    Which model shows how many are in each equal group?

    (A)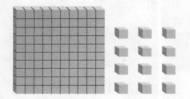

    (C)

    (B)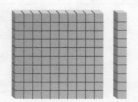

    (D)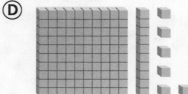

17. Manny has 39 rocks. He wants to put the same number of rocks in each of 7 boxes. Which sentence shows how many rocks will be in each box?

    (A) He will need 6 boxes.

    (B) There will be 6 rocks in each box.

    (C) There will be 5 rocks in each box.

    (D) There will be 5 rocks left over.

Name _____

Fill in the bubble completely to show your answer.

18. There are 176 students in the marching band. They are arranged in equal rows of 8 students for a parade. How many rows of students are there?

Ⓐ 220 rows

Ⓑ 120 rows

Ⓒ 22 rows

Ⓓ 21 rows

19. Naomi wants to plant 387 tulip bulbs in 9 equal rows. She uses division to find the number of tulips in each row. In which place is the first digit of the quotient?

Ⓐ ones

Ⓑ tens

Ⓒ hundreds

Ⓓ thousands

20. Kevin and 2 friends are playing a game of cards. There are 52 cards in the deck to be shared equally. Kevin wants each player to receive the same number of cards. How many cards will each player receive? How many cards will be left over?

Ⓐ 16 cards and 4 cards left over

Ⓑ 17 cards and 1 card left over

Ⓒ 25 cards and 2 cards left over

Ⓓ 26 cards and no cards left over

21. Which number is the quotient?

$1,125 \div 5 =$ ■

Ⓐ 25

Ⓑ 105

Ⓒ 125

Ⓓ 225

22. Mrs. Valdez bought 6 boxes of roses. Each box had 24 roses. She divided all the roses into 9 equal bunches. How many roses were in each bunch? **Explain** how to use a diagram to help solve the problem. Show your diagrams.

_____

_____

▶ **Performance Task**

23. Mr. Owens plans to rent tables for a spaghetti fundraiser. He needs to seat 184 people.

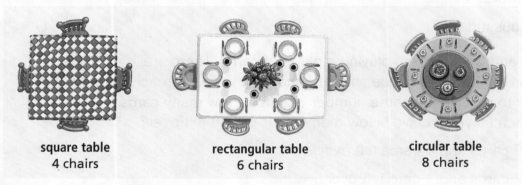

square table
4 chairs

rectangular table
6 chairs

circular table
8 chairs

A If Mr. Owens wants all rectangular tables, how many tables should he rent? **Explain.**

_____

_____

B Square tables rent for $12 each. Circular tables rent for $23 each. Mr. Owens says it would cost him less to rent square tables instead of circular tables. Is he right? **Explain.**

_____

_____

_____

Chapter 5

Factors, Multiples, and Patterns

Show What You Know

Check your understanding of important skills.

Name _____

▶ **Skip-Count** Skip-count to find the unknown numbers.

1. Skip count by 3s.

____3____, _____, _____, _____

2. Skip count by 5s.

____5____, _____, _____, _____

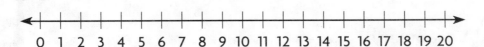

▶ **Arrays** Use the array to find the product.

3.

_____ rows of _____ = _____

4.

_____ rows of _____ = _____

▶ **Multiplication Facts** Find the product.

5. $4 \times 5 =$ _____   **6.** $9 \times 4 =$ _____   **7.** $6 \times 7 =$ _____

MATH DETECTIVE

Recycled plastic helps keep people warm. Some factories use recycled plastic, combined with other fabrics, to make winter jackets. A warehouse has 46 truckloads of recycled plastic. They use 8 truckloads each day. When there are fewer than 16 truckloads, more needs to be ordered. Be a Math Detective. Figure out how many truckloads will be left after 2 days. After 3 days. When will more need to be ordered?

Vocabulary Builder

▶ **Visualize It** •••••••••••••••••••••••••••••

Complete the flow map by using the words with a ✓.

Multiplying

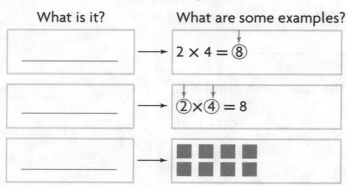

What is it?		What are some examples?
_____	→	$2 \times 4 = ⑧$
_____	→	$②×④ = 8$
_____	→	

▶ **Understand Vocabulary** ••••••••••••••••••••

Complete the sentences by using preview words.

1. A number that is a factor of two or more numbers is a

   _____.

2. A number that is a multiple of two or more numbers is a

   _____.

3. A number that has exactly two factors, 1 and itself, is a

   _____.

4. A number that has more than two factors is a

   _____.

5. A number is _____ by another number if the quotient is a counting number and the remainder is 0.

6. An ordered set of numbers or objects is a

   _____.

7. Each number in a pattern is called a _____.

GO Online • eStudent Edition • Multimedia eGlossary

Name _____

Model Factors

Essential Question How can you use models to find factors?

UNLOCK the Problem REAL WORLD

A **factor** is a number multiplied by another number to find a product. Every whole number greater than 1 has at least two factors, that number and 1.

$18 = 1 \times 18$ $7 = 7 \times 1$ $342 = 1 \times 342$

     ↑   ↑

   factor   factor

Many numbers can be broken into factors in different ways.

$16 = 1 \times 16$ $16 = 4 \times 4$ $16 = 2 \times 8$

Math Idea

When you are asked to find factors of a whole number, only list factors that are also whole numbers.

Activity Model and record the factors of 24.

Materials ■ square tiles

Use all 24 tiles to make as many different arrays as you can. Record the arrays in the grid, and write the factors modeled.

$2 \times 12 = 24$

Factors: _____, _____

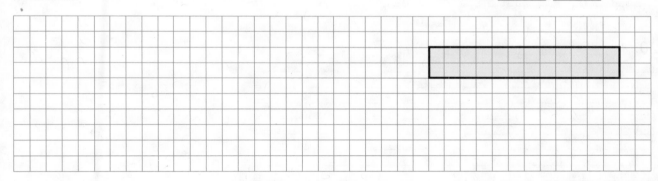

_____ × _____ = 24   _____ × _____ = 24   _____ × _____ = 24

Factors: _____, _____   Factors: _____, _____   Factors: _____, _____

The factors of 24, from least to greatest, are

_____, _____, _____, _____, _____, _____, _____, and _____.

Two factors that make a product are sometimes called a factor pair. How many factor pairs does 24 have? **Explain**.

_____

MATHEMATICAL PRACTICES

Math Talk Can you arrange the tiles in each array another way and show the same factors? **Explain**.

Share and Show

1. Use the arrays to name the factors of 12.

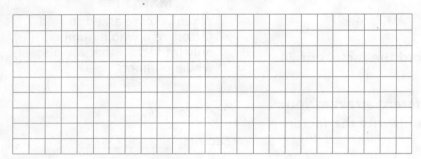

_____ × _____ = 12   _____ × _____ = 12   _____ × _____ = 12

The factors of 12 are 1, _____, 3, _____, 6, and _____.

Use tiles to find all the factors of the product. Record the arrays and write the factors shown.

MATHEMATICAL PRACTICES

Math Talk Explain how the numbers 3 and 12 are related. Use the word *factor* in your explanation.

2. 5: _____

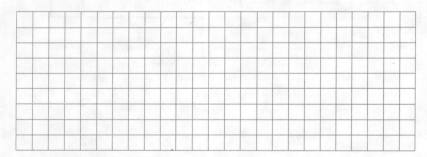

3. 20: _____

4. 25: _____

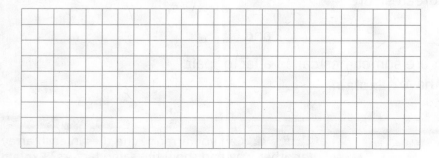

Name _____

On Your Own

Practice: Copy and Solve Use tiles to find all the factors of the product. Record the arrays on grid paper and write the factors shown.

5. 9   **6.** 21   **7.** 17   **8.** 18

Problem Solving · REAL WORLD

Use the diagram for 9–10.

9. **Write Math** ► Pablo is using 36 tiles to make a patio. Can he arrange the tiles in another way and show the same factors? Draw a quick picture and **explain**.

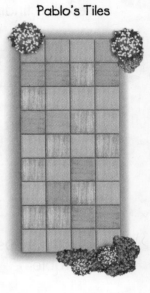

Pablo's Tiles

_____

_____

10. How many different rectangular arrays can Pablo make with all 36 tiles, so none of the arrays show the same factors?

_____

11. **H.O.T.** If 6 is a factor of a number, what other numbers must be factors of the number?

_____

12. Jean spent $16 on new T-shirts. If each shirt cost the same whole-dollar amount, how many could she have bought?

_____

🔑 UNLOCK the Problem REAL WORLD

13. Carmen has 18 connecting cubes. She wants to model a house shaped like a rectangle. If the model has a height of one connecting cube, how many different ways can Carmen model the house using all 18 connecting cubes?

a. What do you need to know? _____

_____

b. How is finding the number of ways to model a rectangular house

related to finding factor pairs? _____

_____

_____

c. Why is finding the factor pairs only the first step in solving the problem? _____

_____

_____

d. Show the steps you used to solve the problem.

_____

_____

_____

_____

e. Complete the sentences. Factor pairs for

18 are _____

_____

There are _____ different ways Carmen can arrange the cubes to model the house.

14. How many ways could Carmen make a rectangular house if she used 24 connecting cubes? **Explain.**

_____

_____

_____

15. Test Prep Which of the following shows a factor pair for the number 16?

(A) 2 and 8

(B) 2 and 32

(C) 8 and 16

(D) 16 and 32

FOR MORE PRACTICE:
Standards Practice Book, pp. P97–P98

Name _____

Factors and Divisibility

Essential Question How can you tell whether one number is a factor of another number?

🔑 UNLOCK the Problem · REAL WORLD

Students in Carlo's art class painted 32 square tiles for a mosaic. They will arrange the tiles to make a rectangle. Can the rectangle have 32 tiles arranged into 3 equal rows, without gaps or overlaps?

🔓 One Way Draw a model.

Think: Try to arrange the tiles into 3 equal rows to make a rectangle.

A rectangle _____ have 32 tiles arranged into 3 equal rows.

🔓 Another Way Use division.

If 3 is a factor of 32, then the unknown factor in 3 × ■ = 32 is a whole number.

$3\overline{)3\ 2}$

Think: Divide to see whether the unknown factor is a whole number.

The unknown factor in 3 × ■ = 32 _____ a whole number.

So, a rectangle _____ have 32 tiles arranged in 3 rows.

▲ Mosaics are decorative patterns made with pieces of glass or other materials.

Math Idea

A factor of a number divides the number evenly. This means the quotient is a whole number and the remainder is 0.

Math Talk MATHEMATICAL PRACTICES

Explain how the model relates to the quotient and remainder for 32 ÷ 3.

- **Explain** how you can tell if 4 is a factor of 30.

_____

_____

Divisibility Rules A number is **divisible** by another number if the quotient is a counting number and the remainder is 0.

Some numbers have a divisibility rule. You can use a divisibility rule to tell whether one number is a factor of another.

 Is 6 a factor of 72?

Think: If 72 is divisible by 6, then 6 is a factor of 72.

Test for divisibility by 6:

Is 72 even? _____

What is the sum of the digits of 72?

_____ + _____ = _____

Is the sum of the digits divisible by 3?

_____

72 is divisible by _____.

So, 6 is a factor of 72.

Divisibility Rules	
Number	**Divisibility Rule**
2	The number is even.
3	The sum of the digits is divisible by 3.
5	The last digit is 0 or 5.
6	The number is even and divisible by 3.
9	The sum of the digits is divisible by 9.

Try This! List all the factor pairs for 72 in the table.

Complete the table.

Factors of 72	
1 × 72 = 72	1, 72
_____ × _____ = _____	_____ , _____
_____ × _____ = _____	_____ , _____
_____ × _____ = _____	_____ , _____
_____ × _____ = _____	_____ , _____
_____ × _____ = _____	_____ , _____

Show your work.

- How did you check if 7 is a factor of 72? **Explain.**

_____

_____

Math Talk **MATHEMATICAL PRACTICES** How are divisibility and factors related? **Explain.**

Name _____

Share and Show

1. Is 4 a factor of 28? Draw a model to help.

   Think: Can you make a rectangle with 28 squares in 4 equal rows?

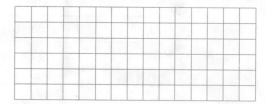

   4 _____ a factor of 28.

Math Talk If 3 is a factor of a number, is 6 always a factor of the number? Explain.

Is 5 a factor of the number? Write *yes* or *no*.

2. 27

☑ 3. 30

4. 36

☑ 5. 53

On Your Own

Is 9 a factor of the number? Write *yes* or *no*.

6. 54

7. 63

8. 67

9. 93

List all the factor pairs in the table.

10.

Factors of 24	
___ × ___ = ___	___ , ___
___ × ___ = ___	___ , ___
___ × ___ = ___	___ , ___
___ × ___ = ___	___ , ___

11.

Factors of 39	
___ × ___ = ___	___ , ___
___ × ___ = ___	___ , ___

Practice: Copy and Solve List all the factor pairs for the number. Make a table to help.

12. 56

13. 64

_____

_____

Problem Solving REAL WORLD

Use the table to solve 14–15.

Stamps Sets	
Country	**Number of stamps**
Germany	90
Sweden	78
Japan	63
Canada	25

14. Dirk bought a set of stamps. The number of stamps in the set he bought is divisible by 2, 3, 5, 6, and 9. Which set is it?

_____

15. **H.O.T.** Geri wants to put 6 stamps on some pages in her stamp book and 9 stamps on other pages. **Explain** how she could do this with the stamp set for Sweden.

_____

_____

_____

_____

_____

SHOW YOUR WORK

16. **H.O.T.** **What's the Error?** George said if 2 and 4 are factors of a number, then 8 is a factor of the number. Is he correct? **Explain.**

_____

_____

_____

17. **Test Prep** Mrs. Mastrioni bought a set of 80 stamps. She wanted to give all the stamps to her students as a reward. She could give equal numbers of stamps to

Ⓐ 2 or 3 students.

Ⓑ 2 or 6 students.

Ⓒ 2, 4, 5, or 8 students.

Ⓓ 2, 4, 8, or 9 students.

FOR MORE PRACTICE:
Standards Practice Book, pp. P99–P100

Name _____

Problem Solving • Common Factors

Essential Question How can you use the *make a list* strategy to solve problems with common factors?

🔑 UNLOCK the Problem · REAL WORLD

Chuck has a coin collection with 30 pennies, 24 quarters, and 36 nickels. He wants to arrange the coins into rows. Each row will have the same number of coins, and all the coins in a row will be the same. How many coins can he put in each row?

The information in the graphic organizer below will help you solve the problem.

Read the Problem	Solve the Problem
What do I need to find?	I can list all the factors of each number. Then I can circle the factors that are common to all three numbers.
I need to find _____ that can go in each row so that each row has _____ _____ .	Factors of: 30 24 36
What information do I need to use?	
Chuck has _____ _____ . Each row has _____ _____ _____ .	
How will I use the information?	
I can make a list to find all the factors of _____ . Then I can use the list to find the common factors. A **common factor** is a factor of two or more numbers.	The common factors are _____ .

So, Chuck can put _____, _____, _____, or _____ coins in each row.

Try Another Problem

Ryan collects animal figures. He has 45 elephants, 36 zebras, and 18 tigers. He will arrange the figures into rows. Each row will have the same number of figures, and all the figures in a row will be the same. How many figures can be in each row?

Use the graphic organizer below to help you solve the problem.

Read the Problem	Solve the Problem
What do I need to find?	
What information do I need to use?	
How will I use the information?	

So, Ryan can put _____, _____, or _____ figures in each row.

Math Talk

How did making a list help you solve the problem?

Name _____

 **Mid-Chapter Checkpoint**

▶ **Vocabulary**

Choose the best term from the box.

1. A number that is multiplied by another number to find a

   product is called a _____. (p. 193)

2. A number is _____ by another number if the
   quotient is a counting number and the remainder is zero. (p. 198)

▶ **Concepts and Skills**

List all the factors from least to greatest.

3. 8

   _____

4. 14

   _____

Is 6 a factor of the number? Write *yes* or *no*.

5. 81   6. 45   7. 42   8. 56

   _____   _____   _____   _____

List all the factor pairs in the table.

9.

Factors of 64	
___ × ___ = ___	___ , ___
___ × ___ = ___	___ , ___
___ × ___ = ___	___ , ___
___ × ___ = ___	___ , ___

10.

Factors of 44	
___ × ___ = ___	___ , ___
___ × ___ = ___	___ , ___
___ × ___ = ___	___ , ___

List the common factors of the numbers.

11. 9 and 18

   _____

12. 20 and 50

   _____

Fill in the bubble completely to show your answer.

13. Sean places 28 tomato plants in rows. All rows contain the same number of plants. Which of these can be the number of plants in a row?

    Ⓐ 3
    Ⓑ 5
    Ⓒ 7
    Ⓓ 8

14. Ella bought some key chains. If she paid $24 for the key chains, and each one cost the same whole-dollar amount, how many could she have bought?

    Ⓐ 5
    Ⓑ 6
    Ⓒ 7
    Ⓓ 10

15. Sandy has 16 roses, 8 daisies, and 32 tulips. She wants to arrange all the flowers in bouquets. Each bouquet has the same number of flowers and the same type of flower. How many flowers could be in a bouquet?

    Ⓐ 3
    Ⓑ 6
    Ⓒ 8
    Ⓓ 16

16. Amir arranged 9 photos on a bulletin board. He put the photos in rows. Each row contains the same number of photos. How many photos could be in each row?

    Ⓐ 1, 3, or 6
    Ⓑ 1, 2, or 9
    Ⓒ 1, 3, or 9
    Ⓓ 3, 6, or 9

Name _____

Factors and Multiples

Essential Question How are factors and multiples related?

🔓 UNLOCK the Problem REAL WORLD

Toy animals are sold in sets of 3, 5, 10, and 12. Mason wants to make a display with 3 animals in each row. Which sets could he buy, if he wants to display all of the animals?

The product of two numbers is a multiple of each number. Factors and multiples are related.

$$3 \times 4 = 12$$

factor factor multiple of 3
              multiple of 4

- How many animals will be in each row?
  _____

- How many animals are sold in each set?
  _____

🔑 One Way Find factors.

Tell whether 3 is a factor of each number.

Think: If a number is divisible by 3, then 3 is a factor of the number.

Is 3 a factor of 3? _____

Is 3 a factor of 5? _____

Is 3 a factor of 10? _____

Is 3 a factor of 12? _____

3 is a factor of _____ and _____.

🔑 Another Way Find multiples.

Multiply and make a list. _3_, _____, _____, _____, _____, ...

    1 × 3   2 × 3   3 × 3   4 × 3   5 × 3

_____ and _____ are multiples of 3.

So, Mason could buy sets of _____ and _____ toy animals.

Math Talk MATHEMATICAL PRACTICES
Explain how you can use what you know about factors to determine whether one number is a multiple of another number.

Common Multiples A **common multiple** is a multiple of two or more numbers.

🔑 **Example** Find common multiples.

Tony works every 3 days and Amanda works every 5 days. If Tony works June 3 and Amanda works June 5, on what days in June will they work together?

Circle multiples of 3. Draw a box around multiples of 5.

June						
Sun	Mon	Tue	Wed	Thu	Fri	Sat
	1	2	3	4	5	6
7	8	9	10	11	12	13
14	15	16	17	18	19	20
21	22	23	24	25	26	27
28	29	30				

Think: The common multiples have both a circle and a box.

The common multiples are _____ and _____.

So, Tony and Amanda will work together on June _____ and June _____.

Share and Show ....

1. Multiply to list the next five multiples of 4.

   ___4___, _____, _____, _____, _____, _____

   1 × 4

Math Talk How are the numbers 5 and 15 related? **Explain.**

MATHEMATICAL PRACTICES

Is the number a factor of 6? Write *yes* or *no*.

✓ 2. 3   3. 6   4. 16   5. 18

   _____   _____   _____   _____

Is the number a multiple of 6? Write *yes* or *no*.

✓ 6. 3   7. 6   8. 16   9. 18

   _____   _____   _____   _____

Name _____

Prime and Composite Numbers

Essential Question How can you tell whether a number is prime or composite?

 UNLOCK the Problem REAL WORLD

Students are arranging square tables to make one larger, rectangular table. If the students want to choose from the greatest number of ways to arrange the tables, should they use 12 or 13 square tables?

Use a grid to show all the possible arrangements of 12 and 13 tables.

Draw all of the possible arrangements of 12 tables and 13 tables. Label each drawing with the factors modeled.

* What are the factors of 12?

_____

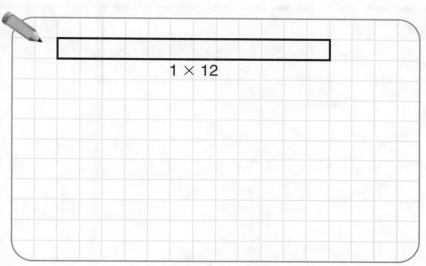

1×12

 ERROR Alert

The same factors in a different order should be counted only once. For example, 3×4 and 4×3 are the same factor pair.

So, there are more ways to arrange _____ tables.

Math Talk MATHEMATICAL PRACTICES
Explain how knowing whether 12 and 13 are prime or composite could have helped you solve the problem above.

* A **prime number** is a whole number greater than 1 that has exactly two factors, 1 and itself.

* A **composite number** is a whole number greater than 1 that has more than two factors.

Factors of 12: _____ , _____ , _____ , _____ , _____ , _____

Factors of 13: _____ , _____

12 is a _____ number, and 13 is a _____ number.

Divisibility You can use divisibility rules to help tell whether a number is prime or composite. If a number is divisible by any number other than 1 and itself, then the number is composite.

 Tell whether 51 is *prime* or *composite*.

Is 51 divisible by 2?

_____

Is 51 divisible by 3?

_____

Think: 51 is divisible by a number other than 1 and 51.
51 has more than two factors.

So, 51 is _____.

> ### Math Idea
> The number 1 is neither prime nor composite, since it has only one factor: 1.

Share and Show MATH BOARD . . . .

1. Use the grid to model the factors of 18. Tell whether 18 is *prime* or *composite*.

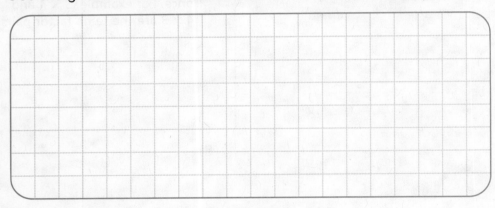

Factors of 18: _____, _____, _____, _____, _____, _____

Think: 18 has more than two factors.

So, 18 is _____.

Tell whether the number is *prime* or *composite*.

2. 11
   Think: Does 11 have other factors besides 1 and itself?

   _____

3. 73

   _____

4. 69

   _____

5. 42

   _____

MATHEMATICAL PRACTICES

Math Talk Is the product of two prime numbers prime or composite? **Explain.**

On Your Own ....

Tell whether the number is *prime* or *composite*.

6. 18 _____

7. 49 _____

8. 29 _____

9. 64 _____

10. 33 _____

11. 89 _____

12. 52 _____

13. 76 _____

H.O.T. **Write *true* or *false* for each statement. Explain or give an example to support your answer.**

14. The number 1 is not prime.

_____

_____

_____

15. A composite number cannot have three factors.

_____

_____

_____

16. Only odd numbers are prime numbers.

_____

_____

17. Every multiple of 7 is a composite number.

_____

_____

Problem Solving ....

18. Name a 2-digit odd number that is prime. Name a 2-digit odd number that is composite.

_____

_____

19. **Test Prep** The number 2 is

Ⓐ prime

Ⓑ composite

Ⓒ neither prime nor composite

Ⓓ both prime and composite

Connect to Social Studies

The Sieve of Eratosthenes

Eratosthenes was a Greek mathematician who lived more than 2,200 years ago. He invented a method of finding prime numbers, which is now called the Sieve of Eratosthenes.

20. Follow the steps below to circle all prime numbers less than 100. Then list the prime numbers.

STEP 1	**STEP 2**	**STEP 3**	**STEP 4**
Cross out 1, since 1 is not prime	Circle 2, since it is prime. Cross out all other multiples of 2.	Circle the next number that is not crossed out. This number is prime. Cross out all the multiples of this number.	Repeat Step 3 until every number is either circled or crossed out.

1	2	3	4	5	6	7	8	9	10
11	12	13	14	15	16	17	18	19	20
21	22	23	24	25	26	27	28	29	30
31	32	33	34	35	36	37	38	39	40
41	42	43	44	45	46	47	48	49	50
51	52	53	54	55	56	57	58	59	60
61	62	63	64	65	66	67	68	69	70
71	72	73	74	75	76	77	78	79	80
81	82	83	84	85	86	87	88	89	90
91	92	93	94	95	96	97	98	99	100

So, the prime numbers less than 100 are

_____

_____

_____

_____

_____

_____

21. Explain why the multiples of any number other than 1 are not prime numbers.

_____

© Houghton Mifflin Harcourt Publishing Company

Name _____

Number Patterns

Essential Question How can you make and describe patterns?

🔑 UNLOCK the Problem · REAL · WORLD

Daryl is making a pattern for a quilt. The pattern shows 40 squares. Every fourth square is blue. How many blue squares are in the pattern?

A **pattern** is an ordered set of numbers or objects. Each number or object in the pattern is called a **term**.

- Underline what you are asked to find.
- Circle what you need to use.

Activity Find a pattern.

Materials ▪ color pencils

Shade the squares that are blue.

1	2	3	4	5	6	7	8	9	10
11	12	13	14	15	16	17	18	19	20
21	22	23	24	25	26	27	28	29	30
31	32	33	34	35	36	37	38	39	40

Which squares are blue? _____

So, there are _____ blue squares in the pattern.

Math Talk MATHEMATICAL PRACTICES
Describe another number pattern in Daryl's quilt.

---

1. What patterns do you see in the arrangement of the blue squares?

   _____

   _____

2. What patterns do you see in the numbers of the blue squares?

   _____

   _____

🔑 Example Find and describe a pattern.

The rule for the pattern is *add* 5. The first term in the pattern is 5.

Ⓐ Use the rule to write the numbers in the pattern.

5   10

5, 10, _____, _____, _____, _____, _____, _____, _____, ...

Ⓑ Describe other patterns in the numbers.

What do you notice about the digits in the ones place?

_____

_____

Describe the pattern using the words *odd* and *even*.

_____

_____

Describe the pattern using the word *multiples*.

_____

_____

Try This! Find and describe a pattern.

The rule for the pattern is *add* 3, *subtract* 1. The first term in the pattern is 6.

Add 3.   Subtract 1.   Add 3.

6

___ ___ ___   ___ ___   ___ ___   ___ ___   ___

Describe another pattern in the numbers.

_____

_____

_____

Name _____

Share and Show . . . .

Use the rule to write the numbers in the pattern.

Math Talk MATHEMATICAL PRACTICES
Explain how the first term in a pattern helps you find the next term.

1. Rule: Subtract 10.   First term: 100

    Think Subtract 10

    100

    _____ _____ _____ _____

    100, _____, _____, _____, _____, …

Use the rule to write the numbers in the pattern.
Describe another pattern in the numbers.

2. Rule: Multiply by 2.   First term: 4

    4, _____, _____, _____, _____, …

    _____

3. Rule: Skip-count by 6.   First term: 12

    12, _____, _____, _____, _____, …

    _____

On Your Own . . . .

Use the rule to write the first twelve numbers in the pattern.
Describe another pattern in the numbers.

4. Rule: Add 7.   First term: 3

    _____

    _____

5. Rule: Subtract 5.   First term: 94

    _____

    _____

6. Rule: Subtract 2, add 3.   First term: 6

    _____

    _____

7. Rule: Add 2, add 1.   First term: 12

    _____

    _____

Problem Solving REAL WORLD

8. The odd- and even-numbered hotel rooms are on different sides of the hall. Room 231 is between which two rooms?

_____

_____

9. Test Prep Which pattern follows the rule _add_ 3, _subtract_ 1?

Ⓐ 60, 63, 60, 63, …

Ⓑ 3, 1, 4, 2, …

Ⓒ 60, 63, 62, 65, …

Ⓓ 60, 63, 66, 69, …

Pose a Problem

10. H.O.T. An activity at the Math Fair shows two charts.

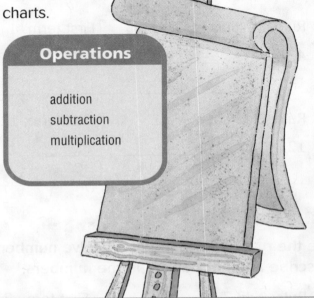

Numbers
2
3
5
6
10

Operations
addition
subtraction
multiplication

Use at least two of the numbers and an operation from the charts to write a pattern problem. Include the first five terms of your pattern in the solution to your problem.

Pose a problem.	Solve your problem.

• **Describe** other patterns in the terms you wrote.

_____

_____

✓ Chapter Review/Test

▶ Vocabulary

Choose the best term from the box.

Vocabulary
composite number
factor
multiple
prime number

1. The product of two numbers is a _____ of both numbers. (p. 207)

2. A _____ has exactly two factors. (p. 211)

3. A number is always a multiple of its _____. (p. 207)

▶ Concepts and Skills

List all the factor pairs in the table.

4.

Factors of 48	
_____ × _____ = _____	_____ , _____
_____ × _____ = _____	_____ , _____
_____ × _____ = _____	_____ , _____
_____ × _____ = _____	_____ , _____
_____ × _____ = _____	_____ , _____

5.

Factors of 81	
_____ × _____ = _____	_____ , _____
_____ × _____ = _____	_____ , _____
_____ × _____ = _____	_____ , _____

Is the number a multiple of 9? Write *yes* or *no*.

6. 3 _____   7. 39 _____   8. 45 _____   9. 93 _____

Tell whether the number is *prime* or *composite*.

10. 65 _____   11. 37 _____   12. 77 _____

Use the rule to write the first twelve terms in the pattern.
Describe another pattern in the numbers.

13. Rule: Add 10, subtract 5. _____

First term: 11 _____

GO Online Assessment Options **Chapter Test**

Fill in the bubble completely to show your answer.

14. Erica knits 18 squares on Monday. She knits 7 more squares each day for the rest of the week. How many squares does Erica have on Friday?

Ⓐ 36

Ⓑ 46

Ⓒ 54

Ⓓ 90

15. James works in a flower shop. He will put 36 tulips in vases for a wedding. He must use the same number of tulips in each vase. How many tulips could be in each vase?

Ⓐ 1, 2, 8

Ⓑ 2, 4, 8

Ⓒ 2, 4, 9

Ⓓ 6, 12, 16

16. What multiple of 7 is a factor of 7?

Ⓐ 0

Ⓑ 1

Ⓒ 7

Ⓓ 14

17. Hot dogs come in packages of 6. Hot dog buns come in packages of 8. Antonio will buy the same number of hot dogs as hot dog buns. How many hot dogs could he buy?

Ⓐ 6

Ⓑ 8

Ⓒ 18

Ⓓ 24

18. Sean has 54 flower bulbs. He planted all the bulbs in rows. Each row has the same number of bulbs. How many bulbs could be in each row?

Ⓐ 6   Ⓑ 8

Ⓒ 12   Ⓓ 26

19. An ice-cream truck visits Julio's street every 3 days and Lara's street every 4 days. The truck visits both streets on April 12. When will the truck visit both streets next?

April						
Sun	Mon	Tue	Wed	Thu	Fri	Sat
1	2	3	4	5	6	7
8	9	10	11	12	13	14
15	16	17	18	19	20	21
22	23	24	25	26	27	28
29	30					

Ⓐ April 15

Ⓑ April 16

Ⓒ April 19

Ⓓ April 24

20. The factors of a number include 2, 3, 4, 6, 8, 12, 16, 32, and 48. Which could be the number?

Ⓐ 32

Ⓑ 64

Ⓒ 96

Ⓓ 98

21. Ms. Booth has 16 red buttons and 24 blue buttons. She is making finger puppets. Each puppet has the same number of blue buttons and red buttons. How many puppets can she make if she uses all of the buttons?

Ⓐ 1, 2, 4, or 8

Ⓑ 1, 2, 4, 8, or 16

Ⓒ 1, 2, 4, 8, or 24

Ⓓ 1, 2, 4, 8, 16, or 24

▶ Constructed Response

22. I am a number between 60 and 100. My ones digit is two less than my tens digit. I am a prime number. What number am I? Explain.

_____

_____

_____

_____

▶ Performance Task

23. The number of pieces on display at an art museum is shown in the table.

Art	
Type of Art	**Number of pieces**
Oil paintings	30
Photographs	24
Sketches	21

Ⓐ The museum's show for July features 30 oil paintings by different artists. All artists show the same number of paintings and each artist shows more than 1 painting. How many artists could be featured in the show?

_____

Ⓑ The museum wants to display all the art pieces in rows. Each row has the same number of pieces and the same type of pieces. How many pieces could be in each row?

_____

Ⓒ The museum alternates between adding 3 new pieces one month and retiring one piece the following month. If the museum starts with 75 pieces and the pattern continues, write the numbers in the pattern for the next 8 months. Describe other patterns in the numbers.

_____

_____

_____

Fractions and Decimals

Developing an understanding of fraction equivalence, addition and subtraction of fractions with like denominators, and multiplication of fractions by whole numbers

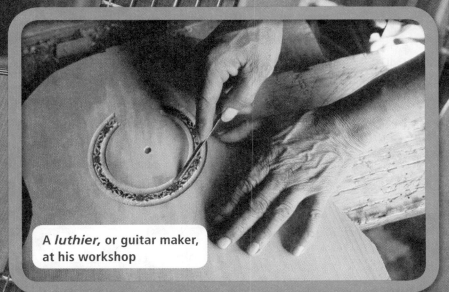

A *luthier,* or guitar maker, at his workshop

Project

Building Custom Guitars

Do you play the guitar, or would you like to learn how to play one? The guitar size you need depends on your height to the nearest inch and on *scale length*. Scale length is the distance from the *bridge* of the guitar to the *nut*.

Get Started

Order the guitar sizes from the least size to the greatest size, and complete the table.

Important Facts

Guitar Sizes for Students			
Age of Player	**Height of Player (to nearest inch)**	**Scale Length (shortest to longest, in inches)**	**Size of Guitar**
4–6	3 feet 3 inches to 3 feet 9 inches	19	
6–8	3 feet 10 inches to 4 feet 5 inches	20.5	
8–11	4 feet 6 inches to 4 feet 11 inches	22.75	
11–Adult	5 feet or taller	25.5	

Size of Guitar: $\frac{1}{2}$ size, $\frac{4}{4}$ size, $\frac{1}{4}$ size, $\frac{3}{4}$ size

Adults play $\frac{4}{4}$-size guitars. You can see that guitars also come in $\frac{3}{4}$, $\frac{1}{2}$, and $\frac{1}{4}$ sizes. Figure out which size guitar you would need according to your height and the scale length for each size guitar. Use the Important Facts to decide. **Explain** your thinking.

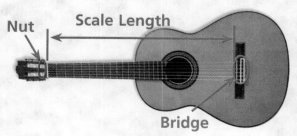

Completed by _____

Fraction Equivalence and Comparison

Show What You Know ✓

Check your understanding of important skills.

Name _____

▶ **Part of a Whole** Write a fraction for the shaded part.

1.  _____

2. _____

3. _____

▶ **Name the Shaded Part** Write a fraction for the shaded part.

4. _____

5. _____

6. _____

▶ **Compare Parts of a Whole** Color the fraction strips to show the fractions. Circle the greater fraction.

7. $\frac{1}{2}$

$\frac{1}{3}$

8. $\frac{1}{5}$

$\frac{1}{3}$

Continent	Part of Land Surface
Asia	$\frac{3}{10}$
Africa	$\frac{1}{5}$
Antarctica	$\frac{9}{100}$
Australia	$\frac{6}{100}$
Europe	$\frac{7}{100}$
North America	$\frac{1}{6}$
South America	$\frac{1}{8}$

Earth's surface is covered by more than 57 million square miles of land. The table shows about how much of Earth's land surface each continent covers. Be a Math Detective. Which continent covers the greatest part of Earth's land surface?

Vocabulary Builder

▶ **Visualize It** ••

Complete the flow map by using the words with a ✓.

Whole Numbers and Fractions

What is it?   What are some examples?

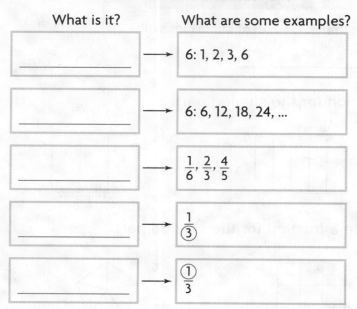

_____ →	6: 1, 2, 3, 6
_____ →	6: 6, 12, 18, 24, ...
_____ →	$\frac{1}{6}, \frac{2}{3}, \frac{4}{5}$
_____ →	$\frac{1}{\textcircled{3}}$
_____ →	$\frac{\textcircled{1}}{3}$

▶ **Understand Vocabulary** •••••••••••••••••••••••••••••••••••

Complete the sentences by using preview words.

1. A fraction is in _____ if the numerator and denominator have only 1 as a common factor.

2. _____ name the same amount.

3. A _____ is a common multiple of two or more denominators.

4. A _____ is a known size or amount that helps you understand a different size or amount.

© Houghton Mifflin Harcourt Publishing Company

Lesson 6.1

Equivalent Fractions

Essential Question How can you use models to show equivalent fractions?

Investigate

Materials ■ color pencils

Joe cut a pan of brownies into third-size pieces. He kept $\frac{1}{3}$ and gave the rest away. Joe will not eat his part all at once. How can he cut his part into smaller, equal-size pieces?

A. Draw on the model to show how Joe could cut his part of the brownies into 2 equal pieces.

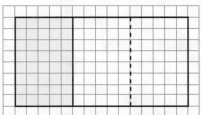

You can rename these 2 equal pieces as a fraction of the original pan of brownies.

Suppose Joe had cut the original pan of brownies into equal pieces of this size.

How many pieces would there be? _____

What fraction of the pan is 1 piece? _____

What fraction of the pan is 2 pieces? _____

You can rename $\frac{1}{3}$ as _____.

B. Now draw on the model to show how Joe could cut his part of the brownies into 4 equal pieces.

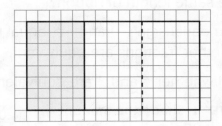

You can rename these 4 equal pieces as a fraction of the original pan of brownies.

Suppose Joe had cut the original pan of brownies into equal pieces of this size.

How many pieces would there be? _____

What fraction of the pan is 1 piece? _____

What fraction of the pan is 4 pieces? _____

You can rename $\frac{1}{3}$ as _____.

C. Fractions that name the same amount are **equivalent fractions**. Write the equivalent fractions.

$$\frac{1}{3} = \frac{}{} = \frac{}{}$$

Draw Conclusions

1. Compare the models for $\frac{1}{3}$ and $\frac{2}{6}$. How does the number of parts relate to the sizes of the parts?

   _____

   _____

2. **Describe** how the numerators are related and how the denominators are related in $\frac{1}{3} = \frac{2}{6}$.

   _____

   _____

3. **H.O.T.** **Apply** Does $\frac{1}{3} = \frac{3}{9}$? **Explain.**

   _____

   _____

Make Connections

Savannah has $\frac{2}{4}$ yard of ribbon, and Lin has $\frac{3}{8}$ yard of ribbon. How can you determine whether Savannah and Lin have the same length of ribbon?

The equal sign (=) and not equal to sign (≠) show whether fractions are equivalent.

Tell whether $\frac{2}{4}$ and $\frac{3}{8}$ are equivalent. Write = or ≠.

STEP 1 Shade the amount of ribbon Savannah has.

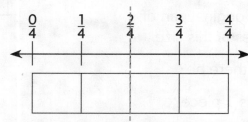

STEP 2 Shade the amount of ribbon Lin has.

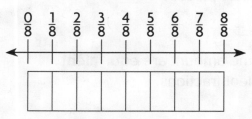

Think: $\frac{2}{4}$ yard is not the same amount as $\frac{3}{8}$ yard.

So, $\frac{2}{4} \bigcirc \frac{3}{8}$.

MATHEMATICAL PRACTICES

Math Talk How could you use a model to show that $\frac{4}{8} = \frac{1}{2}$?

228

© Houghton Mifflin Harcourt Publishing Company

Name _____

Share and Show . . . .

Use the model to write an equivalent fraction.

1.
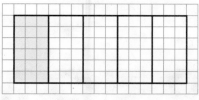

$$\frac{1}{5} \qquad = \qquad \underline{\hspace{2cm}}$$

2.
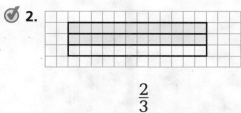

$$\frac{2}{3} \qquad = \qquad \underline{\hspace{2cm}}$$

Tell whether the fractions are equivalent. Write = or ≠.

3. $\frac{1}{6} \bigcirc \frac{2}{12}$

4. $\frac{2}{5} \bigcirc \frac{6}{10}$

5. $\frac{4}{12} \bigcirc \frac{1}{3}$

6. $\frac{5}{8} \bigcirc \frac{2}{4}$

7. $\frac{5}{6} \bigcirc \frac{10}{12}$

8. $\frac{1}{2} \bigcirc \frac{5}{10}$

9. **Write Math** Manny used 8 tenth-size parts to model $\frac{8}{10}$. Ana used fewer parts to model an equivalent fraction. How does the size of a part in Ana's model compare to the size of a tenth-size part? **Explain.**

_____

_____

_____

10. **H.O.T.** How many eighth-size parts do you need to model $\frac{3}{4}$? **Explain.**

_____

_____

11. **Test Prep** Which fraction is equivalent to $\frac{3}{5}$?

A $\frac{6}{8}$

B $\frac{5}{3}$

C $\frac{5}{10}$

D $\frac{6}{10}$

Problem Solving ....

What's the Error?

12. Ben brought two pizzas to a party. He says that since $\frac{1}{4}$ of each pizza is left, the same amount of each pizza is left. What is his error?

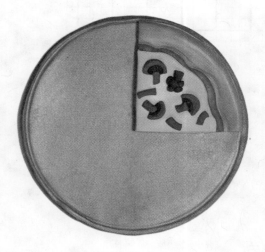

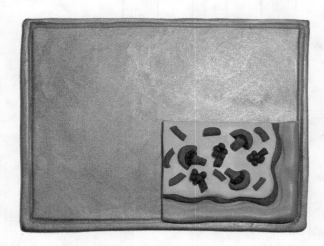

Describe Ben's error.

Draw models of 2 pizzas with a different number of equal pieces. Use shading to show $\frac{1}{4}$ of each pizza.

Name _____

Generate Equivalent Fractions

Essential Question How can you use multiplication to find equivalent fractions?

🔑 UNLOCK the Problem REAL WORLD

Patty needs $\frac{3}{4}$ cup of dish soap to make homemade bubble solution. Her measuring cup is divided into eighths. What fraction of the measuring cup should Patty fill with dish soap?

> • Is an eighth-size part of a measuring cup bigger or smaller than a fourth-size part?
>
> _____

🔒 **Find how many eighths are in $\frac{3}{4}$.**

STEP 1 Compare fourths and eighths.

Shade to model $\frac{1}{4}$.
Use fourth-size parts.

1 part

Shade to model $\frac{1}{4}$.
Use eighth-size parts.

2 parts

You need _____ eighth-size parts to make 1 fourth-size part.

STEP 2 Find how many eighths you need to make 3 fourths.

Shade to model $\frac{3}{4}$.
Use fourth-size parts.

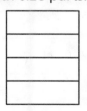

3 parts

Shade to model $\frac{3}{4}$.
Use eighth-size parts.

6 parts

You needed 2 eighth-size parts to make 1 fourth-size part.

So, you need _____ eighth-size parts to make 3 fourth-size parts.

So, Patty should fill $\frac{\quad}{8}$ of the measuring cup with dish soap.

Math Talk ·MATHEMATICAL PRACTICES· How did you know how many eighth-size parts you needed to make 1 fourth-size part? **Explain.**

1. **Explain** why 6 eighth-size parts is the same amount as 3 fourth-size parts.

_____

🔑 Example — Write four fractions that are equivalent to $\frac{1}{2}$.

MODEL	WRITE EQUIVALENT FRACTIONS	RELATE EQUIVALENT FRACTIONS
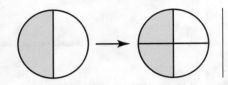	$\frac{1}{2} = \frac{2}{4}$	$\frac{1 \times 2}{2 \times 2} = \frac{2}{4}$
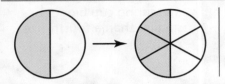	$\frac{1}{2} = \frac{}{6}$	$\frac{1 \times }{2 \times 3} = \frac{}{6}$
	$\frac{1}{2} = \underline{}$	$\frac{1 \times }{2 \times } = \underline{}$
	$\frac{1}{2} = \underline{}$	$\frac{1 \times }{2 \times } = \underline{}$

So, $\frac{1}{2} = \frac{2}{4} = \frac{}{6} = \underline{} = \underline{}$.

2. Look at the model that shows $\frac{1}{2} = \frac{3}{6}$. How does the number of parts in the whole affect the number of parts that are shaded? **Explain**.

_____

_____

3. **Explain** how you can use multiplication to write a fraction that is equivalent to $\frac{3}{5}$.

_____

_____

_____

4. Are $\frac{2}{3}$ and $\frac{6}{8}$ equivalent? **Explain**.

_____

_____

Name _____

Share and Show

1. Complete the table below.

MODEL	WRITE EQUIVALENT FRACTIONS	RELATE EQUIVALENT FRACTIONS
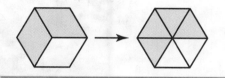	$\frac{2}{3} = \frac{4}{6}$	$\dfrac{2 \times }{3 \times } = \underline{}$
	$\frac{3}{5} = \frac{6}{10}$	$\dfrac{3 \times }{5 \times } = \underline{}$
	$\frac{1}{3} = \frac{4}{12}$	$\dfrac{1 \times }{3 \times } = \underline{}$

> **Math Talk** MATHEMATICAL PRACTICES
> Can you multiply the numerator and denominator of a fraction by 0? **Explain**.

Write two equivalent fractions.

2. $\frac{4}{5}$

$$\frac{4}{5} = \frac{4 \times }{5 \times } = \underline{}$$

$$\frac{4}{5} = \frac{4 \times }{5 \times } = \underline{}$$

$$\frac{4}{5} = \underline{} = \underline{}$$

3. $\frac{2}{4}$

$$\frac{2}{4} = \frac{2 \times }{4 \times } = \underline{}$$

$$\frac{2}{4} = \frac{2 \times }{4 \times } = \underline{}$$

$$\frac{2}{4} = \underline{} = \underline{}$$

On Your Own

Write two equivalent fractions.

4. $\frac{3}{6}$

$$\frac{3}{6} = \underline{} = \underline{}$$

5. $\frac{3}{10}$

$$\frac{3}{10} = \underline{} = \underline{}$$

6. $\frac{2}{5}$

$$\frac{2}{5} = \underline{} = \underline{}$$

Tell whether the fractions are equivalent. Write = or ≠.

7. $\frac{5}{6} \bigcirc \frac{10}{18}$

8. $\frac{4}{5} \bigcirc \frac{8}{10}$

9. $\frac{1}{5} \bigcirc \frac{4}{10}$

10. $\frac{1}{4} \bigcirc \frac{2}{8}$

Problem Solving 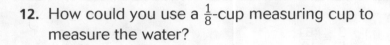 REAL WORLD

Use the recipe for 11–13.

11. How could you use a $\frac{1}{8}$-cup measuring cup to measure the cornstarch?

_____

_____

12. How could you use a $\frac{1}{8}$-cup measuring cup to measure the water?

_____

_____

> **Face Paint Recipe**
>
> $\frac{2}{8}$ cup cornstarch
>
> 1 tablespoon flour
>
> $\frac{9}{12}$ cup light corn syrup
>
> $\frac{1}{4}$ cup water
>
> $\frac{1}{2}$ teaspoon food coloring

13. **H.O.T.** Kim says the amount of flour in the recipe can be expressed as a fraction. Is she correct? **Explain**.

_____

_____

_____

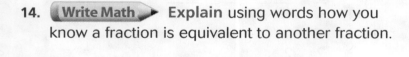

14. **Write Math** ▶ **Explain** using words how you know a fraction is equivalent to another fraction.

_____

_____

_____

_____

_____

15. **Test Prep** Raul needs a piece of rope $\frac{2}{3}$ yard long. Which fraction is equivalent to $\frac{2}{3}$?

(A) $\frac{8}{15}$ yard **(C)** $\frac{8}{12}$ yard

(B) $\frac{6}{12}$ yard **(D)** $\frac{4}{5}$ yard

Name _____

Simplest Form

Essential Question How can you write a fraction as an equivalent fraction in simplest form?

 **UNLOCK the Problem** REAL WORLD

Vicki bought an ice cream cake cut into 6 equal pieces. Vicki, Margo, and Elena each took 2 pieces of the cake home. Vicki says she and each of her friends took $\frac{1}{3}$ of the cake home. Is Vicki correct?

Activity

Materials ■ color pencils

STEP 1 Use a blue pencil to shade the pieces Vicki took home.

_____

STEP 2 Use a red pencil to shade the pieces Margo took home.

_____

STEP 3 Use a yellow pencil to shade the pieces Elena took home.

The cake is divided into _____ equal-size pieces. The 3 colors on the model show how to combine sixth-size pieces to make

_____ equal third-size pieces.

So, Vicki is correct. Vicki, Margo, and Elena each took —— of the cake home.

- Into how many pieces was the cake cut?

_____

- How many pieces did each girl take?

_____

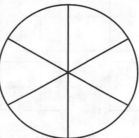

Math Talk MATHEMATICAL PRACTICES
Compare the models for $\frac{2}{6}$ and $\frac{1}{3}$. **Explain** how the sizes of the parts are related.

- **What if** Vicki took 3 pieces of cake home and Elena took 3 pieces of cake home. How could you combine the pieces to write a fraction that represents the part each friend took home? **Explain**.

_____

_____

_____

Simplest Form A fraction is in **simplest form** when you can represent it using as few equal parts of a whole as possible. You need to describe the part you have in equal-size parts. If you can't describe the part you have using fewer parts, then you cannot simplify the fraction.

🔒 One Way Use models to write an equivalent fraction in simplest form.

MODEL	WRITE EQUIVALENT FRACTIONS	RELATE EQUIVALENT FRACTIONS
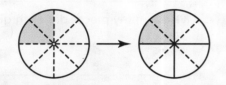	$\dfrac{2}{8} = \dfrac{1}{4}$	$\dfrac{2 \div 2}{8 \div 2} = \dfrac{1}{4}$
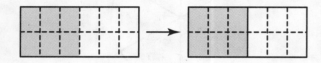	$\dfrac{6}{10} = \dfrac{\blacksquare}{5}$	$\dfrac{6 \div \blacksquare}{10 \div \blacksquare} = \dfrac{\blacksquare}{5}$
	$\dfrac{6}{12} = \dfrac{}{}$	$\dfrac{6 \div \blacksquare}{12 \div \blacksquare} = \dfrac{}{}$

To simplify $\frac{6}{10}$, you can combine tenth-size parts into equal groups with 2 parts each.

So, $\dfrac{6}{10} = \dfrac{6 \div \blacksquare}{10 \div \blacksquare} = \dfrac{}{}$.

🔒 Another Way Use common factors to write $\frac{6}{10}$ in simplest form.

A fraction is in simplest form when 1 is the only factor that the numerator and denominator have in common. The parts of the whole cannot be combined into fewer equal-size parts to show the same fraction.

STEP 1 List the factors of the numerator and denominator. Circle common factors.

Factors of 6: _____, _____, _____, _____

Factors of 10: _____, _____, _____, _____

STEP 2 Divide the numerator and denominator by a common factor greater than 1.

$\dfrac{6}{10} = \dfrac{6 \div \blacksquare}{10 \div \blacksquare} = \dfrac{\blacksquare}{\blacksquare}$

Since 1 is the only factor that 3 and 5 have in common, _____ is written in simplest form.

Name _____

Share and Show

1. Write $\frac{8}{10}$ in simplest form.

$$\frac{8}{10} = \frac{8 \div \boxed{}}{10 \div \boxed{}} = \boxed{}$$

Write the fraction in simplest form.

2. $\frac{6}{12}$

3. $\frac{2}{10}$

4. $\frac{6}{8}$

5. $\frac{4}{6}$

> **Math Talk** MATHEMATICAL PRACTICES
> **Explain** how you know a fraction is in simplest form.

On Your Own

Write the fraction in simplest form.

6. $\frac{9}{12}$

7. $\frac{4}{8}$

8. $\frac{10}{12}$

9. $\frac{20}{100}$

Tell whether the fraction is in simplest form.
Write yes or no.

10. $\frac{2}{8}$

11. $\frac{9}{12}$

12. $\frac{5}{6}$

13. $\frac{4}{10}$

Tell whether the fractions are equivalent.
Write = or ≠. Use simplest form to help.

14. $\frac{3}{6}$ ◯ $\frac{5}{10}$

15. $\frac{9}{12}$ ◯ $\frac{1}{3}$

16. $\frac{3}{12}$ ◯ $\frac{2}{4}$

17. $\frac{6}{8}$ ◯ $\frac{9}{12}$

© Houghton Mifflin Harcourt Publishing Company

Problem Solving REAL WORLD

Use the map for 18–19.

18. What fraction of the states in the southwest region share a border with Mexico? Is this fraction in simplest form?

_____

19. **H.O.T.** **What's the Question?** $\frac{1}{3}$ of the states in this region are on the Gulf of Mexico.

_____

_____

_____

Regions of the United States

- ☐ Northeast
- ☐ Southeast
- ☐ Middle West
- ☐ Southwest
- ☐ West

20. **Sense or Nonsense?** Pete says that to write $\frac{4}{6}$ as $\frac{2}{3}$, you combine pieces, but to write $\frac{4}{6}$ as $\frac{8}{12}$, you break apart pieces. Does this make sense? **Explain**.

_____

_____

_____

_____

.... **SHOW YOUR WORK** ....

21. **Test Prep** Eight of 12 students bought lunch today. In simplest form, what fraction of the students bought lunch today?

Ⓐ $\frac{4}{6}$

Ⓑ $\frac{2}{3}$

Ⓒ $\frac{10}{12}$

Ⓓ $\frac{12}{12}$

© Houghton Mifflin Harcourt Publishing Company

FOR MORE PRACTICE:
Standards Practice Book, pp. P117–P118

Name _____

Common Denominators

Essential Question How can you write a pair of fractions as fractions with a common denominator?

🔓 UNLOCK the Problem · REAL WORLD

Martin has two cakes that are the same size. One cake is cut into $\frac{1}{2}$-size pieces. The other cake is cut into $\frac{1}{3}$-size pieces. He wants to cut the cakes so they have the same size pieces. How can he cut each cake?

A **common denominator** is a common multiple of the denominators of two or more fractions. Fractions with common denominators represent wholes cut into the same number of pieces.

🔓 Activity Use paper folding and shading.

Materials ■ 2 sheets of paper

Find a common denominator for $\frac{1}{2}$ and $\frac{1}{3}$.

STEP 1

Model the cake cut into $\frac{1}{2}$-size pieces. Fold one sheet of paper in half. Draw a line on the fold.

_____

STEP 2

Model the cake cut into $\frac{1}{3}$-size pieces. Fold the other sheet of paper into thirds. Draw lines on the folds.

_____

STEP 3

Fold each sheet of paper so that both sheets have the same number of parts. Draw lines on the folds. How many equal

parts does each sheet of paper have? _____

STEP 4

Draw a picture of your sheets of paper to show how many pieces each cake could have.

So, each cake could be cut into _____ pieces.

> **Math Talk** MATHEMATICAL PRACTICES
> Does Martin need to cut each cake the same number of times? Explain.

🔑 Example Write $\frac{4}{5}$ and $\frac{1}{2}$ as a pair of fractions with common denominators.

You can use common multiples to find a common denominator. List multiples of each denominator. A common multiple can be used as a common denominator.

STEP 1 List multiples of 5 and 2.
Circle common multiples.

5:      5,   10,   _____, _____, _____, _____

2:      _____, _____, _____, _____, _____, _____

STEP 2 Write equivalent fractions.

$$\frac{4}{5} = \frac{4 \times \_\_\_}{5 \times \_\_\_} = \frac{}{10}$$

$$\frac{1}{2} = \frac{1 \times \_\_\_}{2 \times \_\_\_} = \frac{}{10}$$

 Choose a denominator that is a common multiple of 5 and 2.

You can write $\frac{4}{5}$ and $\frac{1}{2}$ as _____ and _____.

> ❗ **ERROR Alert**
>
> Remember that when you multiply the denominator by a factor, you must multiply the numerator by the same factor to write an equivalent fraction.

1. Are $\frac{4}{5}$ and $\frac{1}{2}$ equivalent? **Explain.**

_____

_____

_____

2. Describe another way you could tell whether $\frac{4}{5}$ and $\frac{1}{2}$ are equivalent.

_____

_____

_____

Share and Show ·······························

1. Find a common denominator for $\frac{1}{3}$ and $\frac{1}{12}$ by dividing each whole into the same number of equal parts. Use the models to help.

common denominator: _____

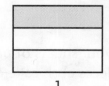 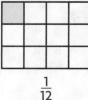

$\frac{1}{3}$ $\frac{1}{12}$

Name _____

Write the pair of fractions as a pair of fractions with
a common denominator.

2. $\frac{1}{2}$ and $\frac{1}{4}$

3. $\frac{3}{4}$ and $\frac{5}{8}$

✓ 4. $\frac{1}{3}$ and $\frac{1}{4}$

✓ 5. $\frac{4}{12}$ and $\frac{5}{8}$

Math Talk MATHEMATICAL PRACTICES
Explain how using a model or listing multiples helps you find a common denominator.

On Your Own

Write the pair of fractions as a pair of fractions with a
common denominator.

6. $\frac{1}{4}$ and $\frac{5}{6}$

7. $\frac{3}{5}$ and $\frac{4}{10}$

Tell whether the fractions are equivalent. Write = or ≠.

8. $\frac{3}{4}$ ◯ $\frac{1}{2}$

9. $\frac{3}{4}$ ◯ $\frac{6}{8}$

10. $\frac{1}{2}$ ◯ $\frac{4}{8}$

11. $\frac{6}{8}$ ◯ $\frac{4}{8}$

12. $\frac{1}{3}$ ◯ $\frac{2}{6}$

13. $\frac{1}{3}$ ◯ $\frac{4}{12}$

14. $\frac{2}{6}$ ◯ $\frac{4}{12}$

15. $\frac{4}{12}$ ◯ $\frac{4}{12}$

Problem Solving  REAL WORLD

16. **Sense or Nonsense?** Carrie has a red streamer that is $\frac{3}{4}$ yard long and a blue streamer that is $\frac{5}{6}$ yard long. She says the streamers are the same length. Does this make sense? **Explain**.

_____

_____

_____

_____

_____

SHOW YOUR WORK

17. **H.O.T.** Leah has two same-size rectangles divided into the same number of equal parts. One rectangle has $\frac{1}{3}$ of the parts shaded, and the other has $\frac{2}{5}$ of the parts shaded. What is the least number of parts into which both rectangles could be divided?

_____

18. **H.O.T. What's the Error?** Jonah says a common denominator for $\frac{3}{4}$ and $\frac{2}{5}$ is 9. What is Jonah's error? **Explain**.

_____

_____

_____

_____

19. **Test Prep** Kevin practiced his trumpet $\frac{2}{3}$ hour on Tuesday and $\frac{3}{4}$ hour on Thursday. Which number below is a common denominator for $\frac{2}{3}$ and $\frac{3}{4}$?

Ⓐ 2   Ⓒ 8

Ⓑ 4   Ⓓ 12

FOR MORE PRACTICE:
Standards Practice Book, pp. P119–P120

Problem Solving • Find Equivalent Fractions

Essential Question How can you use the strategy *make a table* to solve problems using equivalent fractions?

🔑 UNLOCK the Problem · REAL WORLD

Anaya is planting a flower garden. The garden will have no more than 12 equal sections. $\frac{3}{4}$ of the garden will have daisies. What other fractions could represent the part of the garden that will have daisies?

Read the Problem

What do I need to find?	**What information do I need to use?**	**How will I use the information?**
_____ that could represent the part of the garden that will have daisies	_____ of the garden will have daisies. The garden will not have more than _____ equal sections.	I can make a _____ to find _____ fractions to solve the problem.

Solve the Problem

I can make a table and draw models to find equivalent fractions.

$\frac{3}{4}$	$\dfrac{}{}$	$\dfrac{}{}$

1. What other fractions could represent the part of the garden that will have daisies? **Explain.** _____

_____

_____

Math Talk · MATHEMATICAL PRACTICES
Compare the models of the equivalent fractions. How does the number of parts relate to the size of the parts? **Explain.**

🔑 Try Another Problem

Two friends are knitting scarves. Each scarf has 3 rectangles, and $\frac{2}{3}$ of the rectangles have stripes. If the friends are making 10 scarves, how many rectangles do they need? How many rectangles will have stripes?

Read the Problem

What do I need to find?	What information do I need to use?	How will I use the information?

Solve the Problem

2. Does your answer make sense? **Explain** how you know.

_____

_____

_____

© Houghton Mifflin Harcourt Publishing Company

Math Talk MATHEMATICAL PRACTICES
What strategy did you use and why?

Name _____

Share and Show

♀ UNLOCK the Problem `Tips`

✓ Use the Problem Solving Mathboard.
✓ Underline important facts.
✓ Choose a strategy you know.

1. Keisha is helping plan a race route for a 10-kilometer charity run. The committee wants to set up the following things along the course.

> **Viewing areas:** At the end of each half of the course
>
> **Water stations:** At the end of each fifth of the course
>
> **Distance markers:** At the end of each tenth of the course

Which locations have more than one thing located there?

First, make a table to organize the information.

	Number of Locations	First Location	All the Locations
Viewing Areas	2	$\frac{1}{2}$	$\frac{1}{2}$
Water Stations	5	$\frac{1}{5}$	$\frac{1}{5}$
Distance Markers	10	$\frac{1}{10}$	$\frac{1}{10}$

Next, identify a relationship. Use a common denominator, and find equivalent fractions.

Finally, identify the locations at which more than one thing will be set up. Circle the locations.

_____

_____

2. ☀️**H.O.T.** **What if** distance markers will also be placed at the end of every fourth of the course? Will any of those markers be set up at the same location as another distance marker, a water station,

or a viewing area? **Explain.** _____

_____

_____

3. Fifty-six students signed up to volunteer for the race. There were 4 equal groups of students, and each group had a different task.

How many students were in each group? _____

On Your Own

Choose a
STRATEGY

Act It Out
Draw a Diagram
Find a Pattern
Make a Table or List
Solve a Simpler Problem

4. Andy cut a tuna sandwich and a chicken sandwich into a total of 15 same-size pieces. He cut the tuna sandwich into 9 more pieces than the chicken sandwich. Andy ate 8 pieces of the tuna sandwich. What fraction of the tuna sandwich did he eat?

_____

SHOW YOUR WORK

5. A baker cut a pie in half. He cut each half into 3 equal pieces and each piece into 2 equal slices. He sold 6 slices. What fraction of the pie did the baker sell?

_____

6. **H.O.T.** **Write Math** ▸ Luke threw balls into these buckets at a carnival. The number on the bucket gives the number of points for each throw. What is the least number of throws needed to score exactly 100 points? **Explain**.

_____

_____

_____

_____

7. **H.O.T.** A number has exactly eight factors. Two of the factors are 10 and 14. What is the number, and what are all of the factors?

_____

8. **Test Prep** A comic-book store will trade 5 of its comic books for 6 of yours. How many of its comic books will the store trade for 36 of yours?

(A) 25   (B) 30   (C) 36   (D) 42

FOR MORE PRACTICE:
Standards Practice Book, pp. P121–P122

Name _____

 Mid-Chapter Checkpoint

▶ **Vocabulary**

Choose the best term from the box.

1. _____ name the same amount. (p. 227)

2. A _____ is a common multiple of two or more denominators. (p. 239)

▶ **Concepts and Skills**

Write two equivalent fractions.

3. $\frac{2}{5}$ = _____ = _____

4. $\frac{1}{3}$ = _____ = _____

5. $\frac{3}{4}$ = _____ = _____

Tell whether the fractions are equivalent. Write = or ≠.

6. $\frac{2}{3}$ ◯ $\frac{4}{12}$

7. $\frac{5}{6}$ ◯ $\frac{10}{12}$

8. $\frac{1}{4}$ ◯ $\frac{4}{8}$

Write the fraction in simplest form.

9. $\frac{6}{8}$

10. $\frac{25}{100}$

11. $\frac{8}{10}$

Write the pair of fractions as a pair of fractions with a common denominator.

12. $\frac{3}{10}$ and $\frac{2}{5}$

13. $\frac{1}{3}$ and $\frac{3}{4}$

© Houghton Mifflin Harcourt Publishing Company

Fill in the bubble completely to show your answer.

14. Sam needs $\frac{5}{6}$ cup mashed bananas and $\frac{3}{4}$ cup mashed strawberries for a recipe. He wants to find whether he needs more bananas or more strawberries. How can he write $\frac{5}{6}$ and $\frac{3}{4}$ as a pair of fractions with a common denominator?

  (A) $\frac{5}{12}$ and $\frac{3}{12}$

  (B) $\frac{15}{18}$ and $\frac{15}{20}$

  (C) $\frac{10}{12}$ and $\frac{9}{12}$

  (D) $\frac{10}{12}$ and $\frac{18}{24}$

15. Karen will divide her garden into equal parts. She will plant corn in $\frac{8}{12}$ of the garden. What is the fewest number of parts she can divide her garden into?

  (A) 1

  (B) 2

  (C) 3

  (D) 4

16. Olivia is making scarves. Each scarf will have 5 rectangles, and $\frac{2}{5}$ of the rectangles will be purple. How many purple rectangles does she need for 3 scarves?

  (A) 5

  (B) 6

  (C) 7

  (D) 10

17. Paul needs to buy $\frac{5}{8}$ pound of peanuts. The scale at the store measures parts of a pound in sixteenths. Which measure is equivalent to $\frac{5}{8}$ pound?

  (A) $\frac{5}{16}$ pound

  (B) $\frac{8}{16}$ pound

  (C) $\frac{10}{16}$ pound

  (D) $\frac{15}{16}$ pound

Name _____

Compare Fractions Using Benchmarks

Essential Question How can you use benchmarks to compare fractions?

🔓 UNLOCK the Problem · REAL WORLD

Zach made a popcorn snack. He mixed $\frac{5}{8}$ gallon of popcorn with $\frac{1}{2}$ gallon of dried apple rings. Did he use more dried apple rings or more popcorn?

🔑 Activity Compare $\frac{5}{8}$ and $\frac{1}{2}$.

Materials ■ fraction strips

Use fraction strips to compare $\frac{5}{8}$ and $\frac{1}{2}$. Record on the model below.

$\frac{1}{2}$	$\frac{1}{2}$	$\frac{1}{2}$

| $\frac{5}{8}$ | $\frac{1}{8}$ | $\frac{1}{8}$ | $\frac{1}{8}$ | $\frac{1}{8}$ | $\frac{1}{8}$ | $\frac{1}{8}$ | $\frac{1}{8}$ | $\frac{1}{8}$ |

$\frac{5}{8}$ $\frac{1}{2}$

So, Zach used more _____.

MATHEMATICAL PRACTICES

Math Talk **Explain** how the number of eighth-size parts in $\frac{5}{8}$ is related to the number of eighth-size parts you need to make $\frac{1}{2}$.

1. Write five fractions equivalent to $\frac{1}{2}$. What is the relationship between the numerator and the denominator of fractions equivalent to $\frac{1}{2}$?

_____

_____

2. How many eighths are equivalent to $\frac{1}{2}$?

_____

3. How can you compare $\frac{5}{8}$ and $\frac{1}{2}$ without using a model?

_____

_____

Benchmarks A **benchmark** is a known size or amount that helps you understand a different size or amount. You can use $\frac{1}{2}$ as a benchmark to help you compare fractions.

🔑 Example Use benchmarks to compare fractions.

A family hiked the same mountain trail. Evie and her father hiked $\frac{5}{12}$ of the trail before they stopped for lunch. Jill and her mother hiked $\frac{9}{10}$ of the trail before they stopped for lunch. Who hiked farther before lunch?

Compare $\frac{5}{12}$ and $\frac{9}{10}$ to the benchmark $\frac{1}{2}$.

STEP 1 Compare $\frac{5}{12}$ to $\frac{1}{2}$.

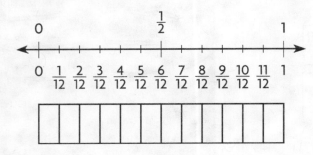

Think: Shade $\frac{5}{12}$.

$\frac{5}{12}$ ◯ $\frac{1}{2}$

STEP 2 Compare $\frac{9}{10}$ to $\frac{1}{2}$.

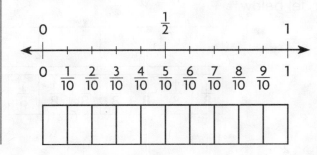

Think: Shade $\frac{9}{10}$.

$\frac{9}{10}$ ◯ $\frac{1}{2}$

Since $\frac{5}{12}$ is _____ than $\frac{1}{2}$ and $\frac{9}{10}$ is _____ than $\frac{1}{2}$, you know that $\frac{5}{12}$ ◯ $\frac{9}{10}$.

So, _____ hiked farther before lunch.

4. Explain how you can tell $\frac{5}{12}$ is less than $\frac{1}{2}$ without using a model.

_____

_____

5. Explain how you can tell $\frac{7}{10}$ is greater than $\frac{1}{2}$ without using a model.

_____

_____

Name _____

Share and Show

1. Compare $\frac{2}{5}$ and $\frac{1}{8}$. Write < or >.

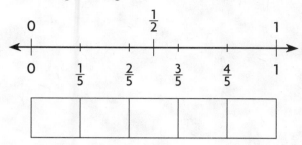
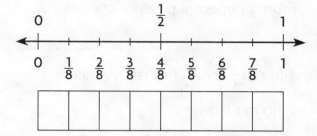

$\frac{2}{5} \bigcirc \frac{1}{8}$

Compare. Write < or >.

2. $\frac{1}{2} \bigcirc \frac{4}{6}$

3. $\frac{3}{10} \bigcirc \frac{1}{2}$

4. $\frac{11}{12} \bigcirc \frac{4}{8}$

5. $\frac{5}{8} \bigcirc \frac{2}{5}$

On Your Own

Math Talk MATHEMATICAL PRACTICES
Explain how you know $\frac{1}{3} < \frac{1}{2}$.

Compare. Write < or >.

6. $\frac{8}{10} \bigcirc \frac{3}{8}$

7. $\frac{1}{3} \bigcirc \frac{7}{12}$

8. $\frac{2}{6} \bigcirc \frac{7}{8}$

9. $\frac{4}{8} \bigcirc \frac{2}{10}$

10. $\frac{3}{4} \bigcirc \frac{1}{2}$

11. $\frac{6}{6} \bigcirc \frac{1}{3}$

12. $\frac{4}{5} \bigcirc \frac{1}{6}$

13. $\frac{5}{8} \bigcirc \frac{9}{10}$

H.O.T. **Algebra** Find a numerator that makes the statement true.

14. $\frac{2}{4} < \frac{}{6}$

15. $\frac{8}{10} > \frac{}{8}$

16. $\frac{10}{12} > \frac{}{4}$

17. $\frac{2}{5} < \frac{}{10}$

18. When two fractions are between 0 and $\frac{1}{2}$, how do you know which fraction is greater? **Explain**.

_____

_____

_____

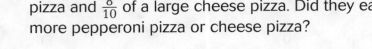

Problem Solving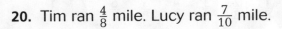

19. A group of students ate $\frac{5}{12}$ of a large pepperoni pizza and $\frac{8}{10}$ of a large cheese pizza. Did they eat more pepperoni pizza or cheese pizza?

_____

20. Tim ran $\frac{4}{8}$ mile. Lucy ran $\frac{7}{10}$ mile.

Who ran farther? _____

21. **H.O.T.** **What's the Question?** Selena ran farther than Manny.

_____

_____

_____

22. Mary made a small pan of ziti and a small pan of lasagna. She cut the ziti into 8 equal parts and the lasagna into 9 equal parts. Her family ate $\frac{2}{3}$ of the lasagna. If her family ate more lasagna than ziti, what fraction of the ziti could have been eaten?

_____

23. **What's the Error?** Tom has two pieces of wood to build a birdhouse. One piece is $\frac{3}{4}$ yard long. The other piece is $\frac{4}{8}$ yard long. Tom says both pieces of wood are the same length. **Explain** his error.

_____

_____

_____

24. **Test Prep** Todd is using the benchmark $\frac{1}{2}$ to compare fractions. Which statement is NOT correct?

Ⓐ $\frac{5}{6} < \frac{1}{2}$　　Ⓒ $\frac{5}{6} > \frac{1}{2}$

Ⓑ $\frac{3}{6} = \frac{1}{2}$　　Ⓓ $\frac{5}{6} \neq \frac{1}{2}$

SHOW YOUR WORK

© Houghton Mifflin Harcourt Publishing Company

Name _____

Compare Fractions

Essential Question How can you compare fractions?

🔓 **UNLOCK the Problem** REAL WORLD

Every year, Avery's school has a fair. This year, $\frac{3}{8}$ of the booths had face painting and $\frac{1}{4}$ of the booths had sand art. Were there more booths with face painting or sand art?

Compare $\frac{3}{8}$ and $\frac{1}{4}$.

🔑 One Way Use a common denominator.

When two fractions have the same denominator, they have equal-size parts. You can compare the number of parts.

THINK

Think: 8 is a multiple of both 4 and 8. Use 8 as a common denominator.

$$\frac{1}{4} = \frac{1 \times }{4 \times } = \frac{}{8}$$

$\frac{3}{8}$ already has 8 as a denominator.

MODEL AND RECORD

Shade the model. Then compare.

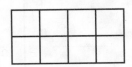 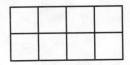

$\frac{3}{8}$ ○ $\frac{2}{8}$

🔑 Another Way Use a common numerator.

When two fractions have the same numerator, they represent the same number of parts. You can compare the size of the parts.

THINK

Think: 3 is a multiple of both 3 and 1. Use 3 as a common numerator.

$\frac{3}{8}$ already has 3 as a numerator.

$$\frac{1}{4} = \frac{1 \times }{4 \times } = \frac{3}{}$$

MODEL AND RECORD

Shade the model. Then compare.

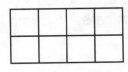 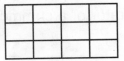

$\frac{3}{8}$ ○ $\frac{3}{12}$

Since $\frac{3}{8}$ ○ $\frac{1}{4}$, there were more booths with _____.

MATHEMATICAL PRACTICES

Math Talk Explain why you cannot use $\frac{1}{2}$ as a benchmark to compare $\frac{3}{8}$ and $\frac{1}{4}$.

Try This! Compare the fractions. Explain your reasoning.

A $\frac{3}{4}$ ◯ $\frac{1}{3}$

_____

_____

_____

_____

_____

B $\frac{3}{5}$ ◯ $\frac{3}{8}$

_____

_____

_____

_____

_____

C $\frac{3}{4}$ ◯ $\frac{7}{8}$

_____

_____

_____

_____

_____

D $\frac{4}{5}$ ◯ $\frac{2}{3}$

_____

_____

_____

_____

_____

1. Which would you use to compare $\frac{11}{12}$ and $\frac{5}{6}$, a common numerator or a common denominator? **Explain**.

_____

_____

2. Can you use simplest form to compare $\frac{8}{10}$ and $\frac{3}{5}$? **Explain**.

_____

_____

254

Name _____

Share and Show

1. Compare $\frac{2}{5}$ and $\frac{1}{10}$.

   Think: Use _____ as a common denominator.

   $\frac{2}{5} = \dfrac{ \times }{ \times } = \underline{}$

   $\frac{1}{10}$

   Think: 4 tenth-size parts $\bigcirc$ 1 tenth-size part.

   $\frac{2}{5} \bigcirc \frac{1}{10}$

2. Compare $\frac{6}{10}$ and $\frac{3}{4}$.

   Think: Use _____ as a common numerator.

   $\frac{6}{10}$

   $\frac{3}{4} = \dfrac{ \times }{ \times } = \underline{}$

   Think: A tenth-size part $\bigcirc$ an eighth-size part.

   $\frac{6}{10} \bigcirc \frac{3}{4}$

Compare. Write <, >, or =.

3. $\frac{7}{8} \bigcirc \frac{2}{8}$ 4. $\frac{5}{12} \bigcirc \frac{3}{6}$ 5. $\frac{4}{10} \bigcirc \frac{4}{6}$ 6. $\frac{6}{12} \bigcirc \frac{2}{4}$

> **Math Talk**   MATHEMATICAL PRACTICES
> **Explain** why using a common numerator or a common denominator can help you compare fractions.

On Your Own

Compare. Write <, >, or =.

7. $\frac{1}{3} \bigcirc \frac{1}{4}$ 8. $\frac{4}{5} \bigcirc \frac{8}{10}$ 9. $\frac{3}{4} \bigcirc \frac{2}{6}$ 10. $\frac{1}{2} \bigcirc \frac{5}{8}$

11. $\frac{3}{10} \bigcirc \frac{2}{4}$ 12. $\frac{75}{100} \bigcirc \frac{8}{10}$ 13. $\frac{4}{6} \bigcirc \frac{2}{3}$ 14. $\frac{3}{10} \bigcirc \frac{4}{100}$

H.O.T. **Algebra** Find a number that makes the statement true.

15. $\frac{1}{2} > \dfrac{}{3}$ 16. $\frac{3}{10} < \dfrac{}{5}$ 17. $\frac{5}{12} < \dfrac{}{3}$ 18. $\frac{2}{3} > \dfrac{4}{}$

© Houghton Mifflin Harcourt Publishing Company

UNLOCK the Problem 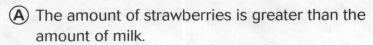 REAL WORLD

19. Jerry is making a strawberry smoothie. Which statement about the recipe is true?

Strawberry Smoothie

3 ice cubes

$\frac{3}{4}$ cup milk

$\frac{2}{6}$ cup cottage cheese

$\frac{8}{12}$ cup strawberries

$\frac{1}{4}$ teaspoon vanilla

$\frac{1}{8}$ teaspoon sugar

  Ⓐ The amount of strawberries is greater than the amount of milk.

  Ⓑ The amount of milk is less than the amount of cottage cheese.

  Ⓒ The amount of strawberries is equal to the amount of cottage cheese.

  Ⓓ The amount of vanilla is greater than the amount of sugar.

a. What do you need to find? _____

b. How will you find the answer? _____

_____

c. Show your work.

d. Fill in the bubble for the correct answer choice above.

20. The number of floats in a parade is a prime number. Which of the following could be the number of floats?

  Ⓐ 16   Ⓒ 41

  Ⓑ 21   Ⓓ 60

21. A kite reached a height of $\frac{4}{12}$ mile. What is $\frac{4}{12}$ in simplest form?

  Ⓐ $\frac{1}{4}$ Ⓒ $\frac{1}{3}$

  Ⓑ $\frac{2}{6}$ Ⓓ $\frac{1}{2}$

FOR MORE PRACTICE:
Standards Practice Book, pp. P125–P126

Name _____

Compare and Order Fractions

Essential Question How can you order fractions?

🔓 UNLOCK the Problem ▷ REAL WORLD ◁

Jody has equal-size bins for the recycling center. She filled $\frac{3}{5}$ of a bin with plastics, $\frac{1}{12}$ of a bin with paper, and $\frac{9}{10}$ of a bin with glass. Which bin is the most full?

- Underline what you need to find.
- Circle the fractions you will compare.

🔑 Example 1 Locate and label $\frac{3}{5}$, $\frac{1}{12}$, and $\frac{9}{10}$ on the number line.

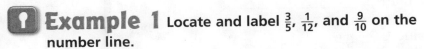

0        $\frac{1}{2}$ 1

Math Idea
Sometimes it is not reasonable to find the exact location of a point on a number line. Benchmarks can help you find approximate locations.

STEP 1 Compare each fraction to $\frac{1}{2}$.

$\frac{3}{5} \bigcirc \frac{1}{2}$ $\frac{1}{12} \bigcirc \frac{1}{2}$ $\frac{9}{10} \bigcirc \frac{1}{2}$

_____ and _____ are both greater than $\frac{1}{2}$.

_____ is less than $\frac{1}{2}$.

Label $\frac{1}{12}$ on the number line above.

STEP 2 Compare $\frac{3}{5}$ and $\frac{9}{10}$.

Think: Use 10 as a common denominator.

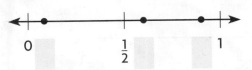

Since $\frac{6}{10} \bigcirc \frac{9}{10}$, you know that $\frac{3}{5} \bigcirc \frac{9}{10}$.

Label $\frac{3}{5}$ and $\frac{9}{10}$ on the number line above.

The fraction the greatest distance from 0 has the greatest value.

The fraction with the greatest value is _____.

So, the bin with _____ is the most full.

Math Talk **Explain** how you know you located $\frac{3}{5}$ on the number line correctly.

MATHEMATICAL PRACTICES

- Compare the distance between $\frac{3}{5}$ and 0 and the distance between $\frac{9}{10}$ and 0. What can you conclude about the relationship between $\frac{3}{5}$ and $\frac{9}{10}$? **Explain**.

_____

_____

© Houghton Mifflin Harcourt Publishing Company

Chapter 6   257

🔒 Example 2 Write $\frac{7}{10}$, $\frac{1}{3}$, $\frac{7}{12}$, and $\frac{8}{10}$ in order from least to greatest.

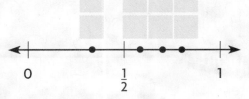

STEP 1 Compare each fraction to $\frac{1}{2}$.

List fractions that are less than $\frac{1}{2}$: _____

List fractions that are greater than $\frac{1}{2}$: _____

The fraction with the least value is _____.

Locate and label $\frac{1}{3}$ on the number line above.
_____

STEP 2 Compare $\frac{7}{10}$ to $\frac{7}{12}$ and $\frac{8}{10}$.

Think: $\frac{7}{10}$ and $\frac{7}{12}$ have a common numerator.

Think: $\frac{7}{10}$ and $\frac{8}{10}$ have a common denominator.

$\frac{7}{10}$ ◯ $\frac{7}{12}$ $\frac{7}{10}$ ◯ $\frac{8}{10}$

Locate and label $\frac{7}{10}$, $\frac{7}{12}$, and $\frac{8}{10}$ on the number line above.

The fractions in order from least to greatest are _____.

So, _____ < _____ < _____ < _____.

Try This! Write $\frac{3}{4}$, $\frac{3}{6}$, $\frac{1}{3}$, and $\frac{2}{12}$ in order from least to greatest.

_____ < _____ < _____ < _____

Name _____

Share and Show

1. Locate and label points on the number line to help you write
$\frac{3}{10}$, $\frac{11}{12}$, and $\frac{5}{8}$ in order from least to greatest.

_____

Write the fraction with the greatest value.

✓ 2. $\frac{7}{10}$, $\frac{1}{5}$, $\frac{9}{10}$

3. $\frac{5}{6}$, $\frac{7}{12}$, $\frac{7}{10}$

4. $\frac{2}{8}$, $\frac{1}{8}$, $\frac{2}{4}$, $\frac{2}{6}$

Write the fractions in order from least to greatest.

✓ 5. $\frac{1}{4}$, $\frac{5}{8}$, $\frac{1}{2}$

6. $\frac{3}{5}$, $\frac{2}{3}$, $\frac{3}{10}$, $\frac{4}{5}$

7. $\frac{3}{4}$, $\frac{7}{12}$, $\frac{5}{12}$

On Your Own

Math Talk MATHEMATICAL PRACTICES
Explain how benchmarks can help you order fractions.

Write the fractions in order from least to greatest.

8. $\frac{2}{5}$, $\frac{1}{3}$, $\frac{5}{6}$

9. $\frac{4}{8}$, $\frac{5}{12}$, $\frac{1}{6}$

10. $\frac{7}{100}$, $\frac{9}{10}$, $\frac{4}{5}$

H.O.T. **Algebra** Write a numerator that makes
the statement true.

11. $\frac{1}{2} < \frac{\boxed{}}{10} < \frac{4}{5}$

12. $\frac{1}{4} < \frac{5}{12} < \frac{\boxed{}}{6}$

13. $\frac{\boxed{}}{8} < \frac{3}{4} < \frac{7}{8}$

🔑 UNLOCK the Problem REAL WORLD

14. Nancy, Lionel, and Mavis ran in a 5-kilometer race. The table shows their finish times. In what order did Nancy, Lionel, and Mavis finish the race?

5-Kilometer Race Results	
Name	**Time**
Nancy	$\frac{2}{3}$ hour
Lionel	$\frac{7}{12}$ hour
Mavis	$\frac{3}{4}$ hour

Finish line

a. What do you need to find?

_____

_____

b. What information do you need to solve the problem?

_____

_____

c. What information is not necessary?

_____

d. How will you solve the problem?

_____

_____

e. Show the steps to solve the problem.

f. Complete the sentences.

The runner who finished first is _____.

The runner who finished second is _____.

The runner who finished third is _____.

15. Alma used 3 beads to make a necklace. The lengths of the beads are $\frac{5}{6}$ inch, $\frac{5}{12}$ inch, and $\frac{1}{3}$ inch. What are the lengths in order from shortest to longest?

16. Test Prep A recipe for Trail Mix includes $\frac{3}{10}$ cup of sunflower seeds, $\frac{1}{2}$ cup of raisins, and $\frac{3}{8}$ cup of granola. Which list shows the amounts from least to greatest?

Ⓐ $\frac{1}{2}$ cup, $\frac{3}{8}$ cup, $\frac{3}{10}$ cup

Ⓑ $\frac{3}{8}$ cup, $\frac{3}{10}$ cup, $\frac{1}{2}$ cup

Ⓒ $\frac{3}{10}$ cup, $\frac{3}{8}$ cup, $\frac{1}{2}$ cup

Ⓓ $\frac{3}{10}$ cup, $\frac{1}{2}$ cup, $\frac{3}{8}$ cup

FOR MORE PRACTICE:
Standards Practice Book, pp. P127–P128

Chapter Review/Test

▶ Vocabulary

Choose the best term from the box.

1. A _____ is a common multiple of two or more denominators. (p. 239)

2. A fraction is in _____ when the numerator and denominator have only 1 as a common factor. (p. 236)

3. A _____ is a known size or amount that helps you understand another size or amount. (p. 250)

▶ Concepts and Skills

Write two equivalent fractions.

4. $\dfrac{4}{6}$ = _____ = _____

5. $\dfrac{6}{10}$ = _____ = _____

6. $\dfrac{2}{8}$ = _____ = _____

Write each pair of fractions as a pair of fractions with a common denominator.

7. $\dfrac{3}{4}$ and $\dfrac{7}{8}$

8. $\dfrac{2}{3}$ and $\dfrac{1}{4}$

9. $\dfrac{7}{10}$ and $\dfrac{4}{5}$

Compare. Write <, >, or =.

10. $\dfrac{5}{8}$ ◯ $\dfrac{5}{12}$

11. $\dfrac{10}{12}$ ◯ $\dfrac{5}{6}$

12. $\dfrac{1}{2}$ ◯ $\dfrac{3}{10}$

13. $\dfrac{1}{4}$ ◯ $\dfrac{1}{3}$

Write the fractions in order from least to greatest.

14. $\dfrac{2}{3}, \dfrac{3}{4}, \dfrac{1}{6}$

15. $\dfrac{7}{10}, \dfrac{4}{5}, \dfrac{1}{2}, \dfrac{4}{12}$

GO Online Assessment Options **Chapter Test**

Fill in the bubble completely to show your answer.

16. Paco needs at least $\frac{3}{8}$ yard of twine to build a model ship. How much twine could he buy?

Ⓐ $\frac{3}{10}$ yard

Ⓑ $\frac{1}{4}$ yard

Ⓒ $\frac{3}{5}$ yard

Ⓓ $\frac{1}{8}$ yard

17. Rachel, Nancy, and Diego were in a fishing competition. Rachel's fish was $\frac{7}{8}$ foot long, Nancy's fish was $\frac{1}{4}$ foot long, and Diego's fish was $\frac{1}{2}$ foot long. What are the lengths of the fish in order from least to greatest?

Ⓐ $\frac{7}{8}$ foot, $\frac{1}{2}$ foot, $\frac{1}{4}$ foot

Ⓑ $\frac{1}{2}$ foot, $\frac{7}{8}$ foot, $\frac{1}{4}$ foot

Ⓒ $\frac{7}{8}$ foot, $\frac{1}{4}$ foot, $\frac{1}{2}$ foot

Ⓓ $\frac{1}{4}$ foot, $\frac{1}{2}$ foot, $\frac{7}{8}$ foot

18. Amy needs $\frac{6}{8}$ gallon of fruit juice to make punch. She needs an equal amount of sparkling water. How much sparkling water does she need?

Ⓐ $\frac{2}{8}$ gallon

Ⓑ $\frac{1}{2}$ gallon

Ⓒ $\frac{2}{3}$ gallon

Ⓓ $\frac{3}{4}$ gallon

19. Gavin is building a model of a kitchen. In the model, $\frac{2}{5}$ of the floor tiles are white, $\frac{1}{2}$ of the floor tiles are yellow, and $\frac{1}{10}$ of the floor tiles are brown. How many floor tiles could be in the model?

Ⓐ 2

Ⓑ 5

Ⓒ 10

Ⓓ 17

Name _____

Fill in the bubble completely to show your answer.

20. Bill has enough money to buy no more than $\frac{1}{2}$ pound of cheese. How much cheese could he buy?

   Ⓐ $\frac{1}{3}$ pound

   Ⓑ $\frac{4}{6}$ pound

   Ⓒ $\frac{5}{8}$ pound

   Ⓓ $\frac{3}{4}$ pound

21. Students planted 6 equal-size gardens on Earth Day. They divided each garden into 3 equal sections and planted herbs in 2 of the 3 sections. What fraction of the gardens did the students plant with herbs?

   Ⓐ $\frac{3}{6}$

   Ⓑ $\frac{2}{6}$

   Ⓒ $\frac{6}{18}$

   Ⓓ $\frac{12}{18}$

22. Noah and Leslie live the same distance from school. Which could be the distances they live from school?

   Ⓐ $\frac{7}{100}$ kilometer and $\frac{7}{10}$ kilometer

   Ⓑ $\frac{5}{10}$ kilometer and $\frac{1}{5}$ kilometer

   Ⓒ $\frac{80}{100}$ kilometer and $\frac{8}{10}$ kilometer

   Ⓓ $\frac{6}{10}$ kilometer and $\frac{2}{5}$ kilometer

23. Keiko needs $\frac{8}{12}$ yard of fabric to finish her quilt. What is $\frac{8}{12}$ written in simplest form?

   Ⓐ $\frac{4}{6}$

   Ⓑ $\frac{2}{3}$

   Ⓒ $\frac{3}{4}$

   Ⓓ $\frac{1}{2}$

Constructed Response

24. Sam needs $\frac{4}{6}$ cup of laundry detergent for his laundry. The cap on top of the laundry detergent holds $\frac{1}{3}$ cup. He has 1 capful of detergent. Does he have enough? **Explain**.

_____

_____

Performance Task

25. The table shows the distances of some places in town from the school.

Distance from School	
Place	**Distance**
Library	$\frac{3}{5}$ mile
Post Office	$\frac{1}{2}$ mile
Park	$\frac{3}{4}$ mile
Town Hall	$\frac{8}{10}$ mile

A Are any of the places shown in the table closer than $\frac{1}{2}$ mile to school? **Explain** how you know.

_____

_____

_____

_____

B Are any of the places shown in the table the same distance from school? **Explain** how you know.

_____

_____

_____

_____

C Which place is farthest from school? **Explain**.

_____

_____

_____

_____

_____

_____

Chapter 7

Add and Subtract Fractions

Show What You Know ✓

Check your understanding of important skills

Name _____

▶ **Fractions Equal to 1** Write the fraction that names the whole.

1. 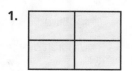 _____

2. 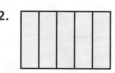 _____

▶ **Parts of a Whole** Write a fraction that names the shaded part.

3. _____

4.  _____

5. 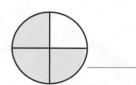 _____

▶ **Read and Write Fractions** Write a fraction for the shaded part. Write a fraction for the unshaded part.

6. (grid)   shaded: _____

   unshaded: _____

7.   shaded: _____

   unshaded: _____

The electricity that powers our appliances is converted from many sources of energy. About $\frac{5}{10}$ is made from coal, about $\frac{2}{10}$ from natural gas, and about $\frac{2}{10}$ from nuclear power. Be a Math Detective. About how much of our electricity comes from sources other than coal, natural gas, or nuclear power?

GO Online   Assessment Options: **Soar to Success Math**

Vocabulary Builder

▶ **Visualize It** ∙∙∙∙∙∙∙∙∙∙∙∙∙∙∙∙∙∙∙∙∙∙∙∙∙∙∙∙∙∙∙∙

Complete the bubble map using the words with a ✓.

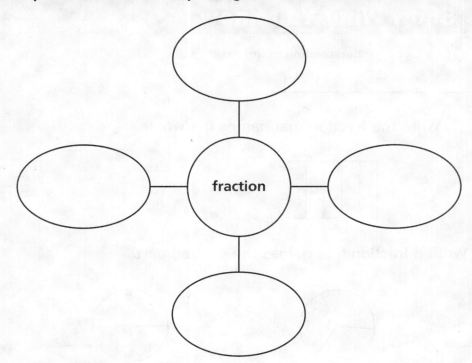

Review Words

Associative Property
   of Addition
Commutative
   Property of
   Addition
✓ denominator
   fraction
✓ numerator
   simplest form

Preview Words

✓ mixed number
✓ unit fraction

▶ **Understand Vocabulary** ∙∙∙∙∙∙∙∙∙∙∙∙∙∙∙∙∙∙∙∙∙∙∙∙∙

Write the word or phrase that matches the description.

1. When the numerator and denominator have only 1 as a common factor

   _____

2. A number that names a part of a whole or part of a group

   _____

3. An amount given as a whole number and a fraction

   _____

4. The number in a fraction that tells how many equal parts

   are in the whole or in the group _____

5. A fraction that has a numerator of one _____

GO Online ∙ eStudent Edition ∙ Multimedia eGlossary

Name _____

Add and Subtract Parts of a Whole

Essential Question When can you add or subtract parts of a whole?

Investigate

Materials ■ fraction circles ■ color pencils

Ms. Clark has the following pie pieces left over from a bake sale.

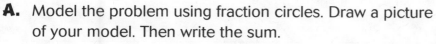

She will combine the pieces so they are on the same dish. How much pie will be on the dish?

A. Model the problem using fraction circles. Draw a picture of your model. Then write the sum.

 + 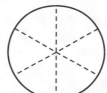 =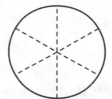

_____ + _____ = _____

So, _____ of a pie is on the dish.

B. Suppose Ms. Clark eats 2 pieces of the pie. How much pie will be left on the dish? Model the problem using fraction circles. Draw a picture of your model. Then write the difference.

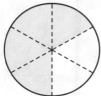

_____ − _____ = _____

So, _____ of the pie is left on the dish.

Draw Conclusions

1. Kevin says that when you combine 3 pieces of pie and 1 piece of pie, you have 4 pieces of pie. **Explain** how Kevin's statement is related to the equation $\frac{3}{6} + \frac{1}{6} = \frac{4}{6}$.

_____

_____

2. Isabel wrote the equation $\frac{1}{2} + \frac{1}{6} = \frac{4}{6}$ and Jonah wrote $\frac{3}{6} + \frac{1}{6} = \frac{4}{6}$ to represent combining the pie pieces. **Explain** why both equations are correct.

_____

3. **H.O.T.** If there is $\frac{4}{6}$ of a pie on a plate, what part of the pie is missing from the plate? Write an equation to justify your answer.

_____

Make Connections

You can only join or separate parts that refer to the same whole.

Suppose Randy has $\frac{1}{4}$ of a round cake and $\frac{1}{4}$ of a square cake.

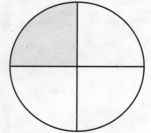 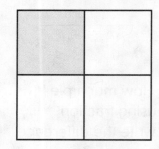

Math Talk
Give an example of a situation where the equation $\frac{1}{4} + \frac{1}{4} = \frac{2}{4}$ makes sense. **Explain** your reasoning.

a. Are the wholes the same? **Explain**.

_____

b. Does the sum $\frac{1}{4} + \frac{1}{4} = \frac{2}{4}$ make sense in this situation? **Explain**.

_____

_____

Name _____

Share and Show

Use the model to write an equation.

1.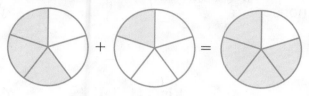

_____

2.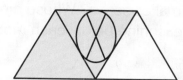

_____

3.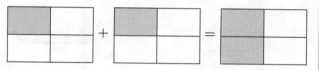

_____

4.

1							
$\frac{1}{8}$	$\frac{1}{8}$	$\frac{1}{8}$	$\frac{1}{8}$	$\frac{1}{8}$	$\frac{1}{8}$	$\frac{1}{8}$	$\frac{1}{8}$

_____

Use the model to solve the equation.

5. $\frac{3}{4} - \frac{1}{4} =$ _____

1			
$\frac{1}{4}$	$\frac{1}{4}$	$\frac{1}{4}$	$\frac{1}{4}$

6. $\frac{5}{6} + \frac{1}{6} =$ _____

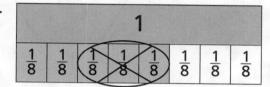

7. Sean has $\frac{1}{5}$ of a cupcake and $\frac{1}{5}$ of a large cake.

   a. Are the wholes the same? **Explain.**

   _____

   b. Does the sum $\frac{1}{5} + \frac{1}{5} = \frac{2}{5}$ make sense in this situation? **Explain.**

   _____

Problem Solving REAL WORLD

Sense or Nonsense?

8. Samantha and Kim used different models to help find $\frac{1}{3} + \frac{1}{6}$. Whose model makes sense? Whose model is nonsense? **Explain** your reasoning below each model.

Samantha's Model

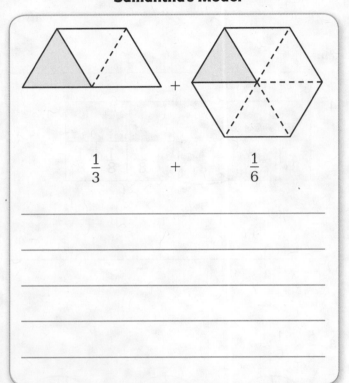

$$\frac{1}{3} \quad + \quad \frac{1}{6}$$

_____

_____

_____

_____

Kim's Model

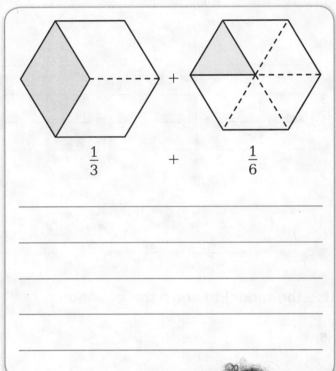

$$\frac{1}{3} \quad + \quad \frac{1}{6}$$

_____

_____

_____

_____

9. Draw a model you could use to add $\frac{1}{4} + \frac{1}{2}$.

FOR MORE PRACTICE:
Standards Practice Book, pp. P133–P134

Write Fractions as Sums

Essential Question How can you write a fraction as a sum of fractions with the same denominators?

🔑 UNLOCK the Problem REAL WORLD

Emilio cut a sandwich into 8 equal pieces and ate 1 piece. He has $\frac{7}{8}$ of the sandwich left. Emilio put each remaining piece on a snack plate. How many snack plates did he use? What part of the sandwich did he put on each plate?

Each piece of the sandwich is $\frac{1}{8}$ of the whole. $\frac{1}{8}$ is called a **unit fraction** because it tells the part of the whole that 1 piece represents. A unit fraction always has a numerator of 1.

🔑 Example 1 Write $\frac{7}{8}$ as a sum of unit fractions.

$\frac{7}{8}$ = _____ + _____ + _____ + _____ + _____ + _____

The number of addends represents the number of plates used.

The unit fractions represent the part of the sandwich on each plate.

So, Emilio used _____ plates. He put _____ of a sandwich on each plate.

1. **What if** Emilio ate 3 pieces of the sandwich instead of 1 piece? How many snack plates would he need? What part of the sandwich would be on each plate? **Explain.**

_____

_____

_____

_____

🔑 **Example 2** Write a fraction as a sum.

Kevin and Olivia are going to share a whole pizza. The pizza is cut into 6 equal slices. They will put the slices on two separate dishes. What part of the whole pizza could be on each dish?

Shade the models to show three different ways Kevin and Olivia could share the pizza. Write an equation for each model.

Think: $\frac{6}{6}$ = 1 whole pizza.

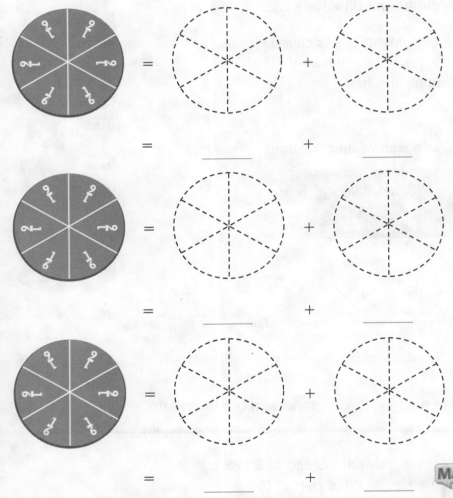

= _____ + _____

= _____ + _____

= _____ + _____

© Houghton Mifflin Harcourt Publishing Company

MATHEMATICAL PRACTICES

Math Talk If there were 8 dishes, could $\frac{1}{6}$ of the whole pizza be on each dish? **Explain.**

2. **What if** 3 friends share the pizza and they put the pizza slices on three separate dishes? What part of the pizza could be on each dish? Write equations to support your answer.

_____

_____

Name _____

Share and Show

1. Write $\frac{3}{4}$ as a sum of unit fractions.

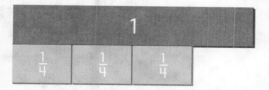

$$\frac{3}{4} = \underline{\hspace{1cm}} + \underline{\hspace{1cm}} + \underline{\hspace{1cm}}$$

Write the fraction as a sum of unit fractions.

✓ 2.

1					
$\frac{1}{6}$	$\frac{1}{6}$	$\frac{1}{6}$	$\frac{1}{6}$	$\frac{1}{6}$	$\frac{1}{6}$

$$\frac{5}{6} = \underline{\hspace{3cm}}$$

✓ 3.

1		
$\frac{1}{3}$	$\frac{1}{3}$	$\frac{1}{3}$

$$\frac{2}{3} = \underline{\hspace{3cm}}$$

On Your Own

Write the fraction as a sum of unit fractions.

4. $\frac{4}{12} = \underline{\hspace{4cm}}$

5. $\frac{6}{8} = \underline{\hspace{4cm}}$

 Math Talk MATHEMATICAL PRACTICES
Explain how the numerator in $\frac{5}{6}$ is related to the number of addends in the sum of its unit fractions.

Write the fraction as a sum of fractions three different ways.

6. $\frac{8}{10}$

_____

_____

_____

7. $\frac{6}{6}$

_____

_____

_____

8. **H.O.T.** How many different ways can you write a fraction that has a numerator of 2 as a sum of fractions? **Explain.**

_____

_____

UNLOCK the Problem · REAL WORLD

9. Holly's garden is divided into 5 equal sections. She will fence the garden into 3 areas by grouping some equal sections together. What part of the garden could each fenced area be?

a. What information do you need to use?

_____

_____

b. How can writing an equation help you solve the problem? _____

_____

_____

c. How can drawing a model help you write an equation?

_____

_____

d. Show how you can solve the problem.

e. Complete the sentence.

The garden can be fenced into _____,

_____, and _____ parts or _____,

_____, and _____ parts.

10. What is $\frac{7}{10}$ written as a sum of unit fractions?

_____

11. **Test Prep** Which is equivalent to $\frac{9}{12}$?

Ⓐ $\frac{5}{12} + \frac{3}{12}$

Ⓑ $\frac{3}{12} + \frac{2}{12} + \frac{1}{12} + \frac{1}{12}$

Ⓒ $\frac{5}{12} + \frac{2}{12} + \frac{2}{12}$

Ⓓ $\frac{4}{12} + \frac{4}{12} + \frac{1}{12} + \frac{1}{12}$

Add Fractions Using Models

Essential Question How can you add fractions with like denominators using models?

🔑 UNLOCK the Problem › REAL WORLD

Ms. Clark made a loaf of bread. She used $\frac{1}{8}$ of the bread for a snack and $\frac{5}{8}$ of the bread for lunch. How much did she use for a snack and lunch?

🔒 One Way Use a picture.

$\frac{1}{8}$ is _____ eighth-size piece of bread.

$\frac{5}{8}$ is _____ eighth-size pieces of bread.

Shade 1 eighth-size piece. Then shade 5 eighth-size pieces.

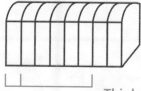

snack lunch

Think: The pieces you shaded represent the pieces Ms. Clark used.

So, Ms. Clark used _____ eighth-size

pieces, or $\frac{\quad}{8}$ of the bread.

🔒 Another Way Use fraction strips.

The 1 strip represents the whole loaf.

Each $\frac{1}{8}$ part represents 1 eighth-size piece of bread.

Shade $\frac{1}{8}$. Then shade $\frac{5}{8}$.

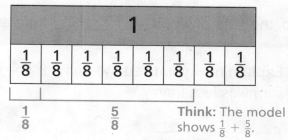

$\frac{1}{8}$ $\frac{5}{8}$

Think: The model shows $\frac{1}{8} + \frac{5}{8}$.

How many $\frac{1}{8}$-size parts are shaded? _____

Write the sum. $\frac{1}{8} + \frac{5}{8} = \frac{\quad}{8}$

So, Ms. Clark used _____ of the bread.

---

1. **Explain** how the numerator of the sum is related to the fraction strip model.

_____

_____

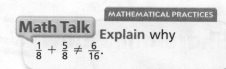

Math Talk MATHEMATICAL PRACTICES
Explain why $\frac{1}{8} + \frac{5}{8} \neq \frac{6}{16}$.

2. **Explain** how the denominator of the sum is related to the fraction strip model.

_____

_____

🔑 Example

Jacob needs two strips of wood to make masts for a miniature sailboat. One mast will be $\frac{3}{6}$ foot long. The other mast will be $\frac{2}{6}$ foot long. He has a strip of wood that is $\frac{4}{6}$ foot long. Is this strip of wood long enough to make both masts?

Shade the model to show $\frac{3}{6} + \frac{2}{6}$.

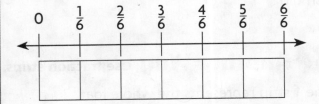

Write the sum. $\frac{3}{6} + \frac{2}{6} = \frac{}{6}$

Is the sum less than or greater than $\frac{4}{6}$? _____

So, the strip of wood _____ long enough to make both masts.

3. **Explain** how you used the number line to determine if the sum was less than $\frac{4}{6}$.

_____

_____

4. **What if** each mast was $\frac{2}{6}$ foot long? Could Jacob use the strip of wood to make both masts? **Explain.**

_____

Share and Show 📝 MATH BOARD . . . .

1. Gary's cat ate $\frac{3}{5}$ of a bag of cat treats in September and $\frac{1}{5}$ of the same bag of cat treats in October. What part of the bag of cat treats did Gary's cat eat in both months?

Use the model to find the sum $\frac{3}{5} + \frac{1}{5}$.

How many fifth-size pieces are shown? _____

$\frac{3}{5} + \frac{1}{5} = \frac{}{5}$ of a bag

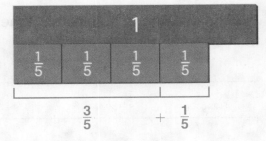

Name _____

Use the model to find the sum.

2.

1			
$\frac{1}{4}$	$\frac{1}{4}$	$\frac{1}{4}$	$\frac{1}{4}$

$\frac{1}{4}$ + $\frac{2}{4}$

$\frac{1}{4} + \frac{2}{4} = $ _____

✅ 3.

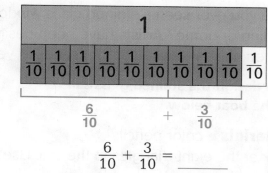

1									
$\frac{1}{10}$	$\frac{1}{10}$	$\frac{1}{10}$	$\frac{1}{10}$	$\frac{1}{10}$	$\frac{1}{10}$	$\frac{1}{10}$	$\frac{1}{10}$	$\frac{1}{10}$	$\frac{1}{10}$

$\frac{6}{10}$ + $\frac{3}{10}$

$\frac{6}{10} + \frac{3}{10} = $ _____

Find the sum. Use models to help.

4. $\frac{3}{6} + \frac{3}{6} = $ _____

✅ 5. $\frac{5}{8} + \frac{2}{8} = $ _____

6. $\frac{1}{3} + \frac{1}{3} = $ _____

On Your Own ....

Find the sum. Use models to help.

Math Talk MATHEMATICAL PRACTICES
Explain how to add $\frac{2}{6} + \frac{3}{6}$.

7. $\frac{5}{8} + \frac{2}{8} = $ _____

8. $\frac{2}{5} + \frac{2}{5} = $ _____

9. $\frac{4}{6} + \frac{1}{6} = $ _____

10. $\frac{1}{10} + \frac{4}{10} = $ _____

11. $\frac{1}{4} + \frac{1}{4} = $ _____

12. $\frac{5}{12} + \frac{5}{12} = $ _____

Problem Solving REAL WORLD

13. **Write Math** Jin is putting colored sand in a jar. She filled $\frac{2}{10}$ of the jar with blue sand and $\frac{4}{10}$ of the jar with pink sand. Describe one way to model the part of the jar filled with sand.

_____

_____

14. **H.O.T.** A sum has five addends. Each addend is a unit fraction. The sum is 1. What are the addends?

_____

15. **Test Prep** Rita needs $\frac{4}{8}$ yard of ribbon to wrap a gift for her sister and $\frac{2}{8}$ yard of ribbon to wrap a gift for a friend. How much ribbon does she need to wrap both gifts?

Ⓐ $\frac{3}{8}$ yard

Ⓑ $\frac{6}{8}$ yard

Ⓒ $\frac{7}{8}$ yard

Ⓓ $\frac{8}{8}$ yard

Connect to Art

Stained Glass Windows

Have you ever seen a stained glass window in a building or home? Artists have been designing stained glass windows for thousands of years.

Help design the stained glass sail on the boat below.

Materials ■ color pencils
Look at the eight triangles in the sail. Use the guide below to color the triangles:

- $\frac{2}{8}$ blue

- $\frac{3}{8}$ red

- $\frac{2}{8}$ orange

- $\frac{1}{8}$ yellow

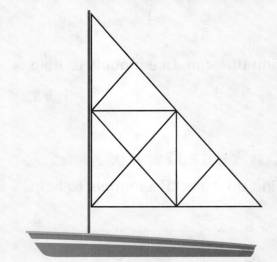

16. Write an equation that shows the fraction of triangles that are red or blue.

_____

17. What part of the sail is orange or yellow? **Explain** how you found the answer.

_____

_____

_____

18. What color is the greatest part of the sail? Write a fraction for that color. How do you know that fraction is greater than the other fractions? **Explain.**

_____

_____

_____

19. (**Write Math**)▶**Pose a Problem** Write a problem about the sail that uses addition of fractions.

_____

_____

_____

FOR MORE PRACTICE:
Standards Practice Book, pp. P137–P138

Subtract Fractions Using Models

Essential Question How can you subtract fractions with like denominators using models?

🔑 UNLOCK the Problem REAL WORLD

A rover needs to travel $\frac{5}{8}$ mile to reach its destination. It has already traveled $\frac{3}{8}$ mile. How much farther does the rover need to travel?

Compare fractions to find the difference.

STEP 1 Shade the model.

Shade the model to show the total distance.

Then shade the model to show how much distance the rover has already covered.

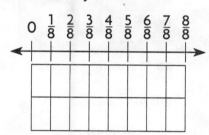

Total distance

Distance traveled

Think: The difference is _____.

STEP 2 Write the difference.

$$\frac{5}{8} - \frac{3}{8} = \frac{}{8}$$

So, the rover needs to travel _____ mile farther.

1. **Explain** how the model shows how much farther the rover needs to travel.

   _____

   _____

2. **Explain** how you can use the model to find $\frac{6}{8} - \frac{2}{8}$.

   _____

   _____

🔑 Example

Sam ordered a small pizza, which was cut into 6 equal slices. He ate $\frac{2}{6}$ of the pizza and put the rest away for later. How much of the pizza did he put away for later?

Find $1 - \frac{2}{6}$.

- How much pizza did Sam begin with?

  _____

- How many slices are in the whole? _____

- How many slices did Sam eat? _____

🔑 One Way Use a picture.

Shade 1 whole.

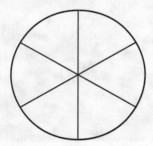

Cross out the parts Sam ate.

Think: He ate $\frac{2}{6}$ of the pizza, or 2 sixth-size parts.

How many sixth-size parts are left? _____

🔑 Another Way Use fraction strips.

Use six $\frac{1}{6}$-size parts to model the whole pizza.

1					
$\frac{1}{6}$	$\frac{1}{6}$	$\frac{1}{6}$	$\frac{1}{6}$	$\frac{1}{6}$	$\frac{1}{6}$

How many $\frac{1}{6}$-size parts should you cross out to model the slices Sam ate? _____

How many $\frac{1}{6}$-size parts are left? _____

Write the difference.

$1 - \dfrac{}{} = \dfrac{}{}$

So, Sam put _____ of the pizza away for later.

Math Talk   MATHEMATICAL PRACTICES
Explain why it makes sense to think of 1 whole as $\frac{6}{6}$ in this problem.

---

3. **Explain** how the equation $\frac{6}{6} - \frac{2}{6} = \frac{4}{6}$ is related to the problem situation.

   _____

   _____

4. 🌟H.O.T.🌟 **Explain** how you could find the unknown addend in $\frac{2}{6} + \blacksquare = 1$ without using a model.

   _____

   _____

   _____

Name _____

Share and Show

1. Lisa needs $\frac{4}{5}$ pound of shrimp to make shrimp salad. She has $\frac{1}{5}$ pound of shrimp. How much more shrimp does Lisa need to make the salad?

   Subtract $\frac{4}{5} - \frac{1}{5}$. Use the model to help.

   Shade the model to show how much shrimp Lisa needs.

   Then shade the model to show how much shrimp Lisa has. Compare the difference between the two shaded rows.

   $\frac{4}{5} - \frac{1}{5} = \frac{}{5}$ pound

   Lisa needs _____ pound more shrimp.

Use the model to find the difference.

2. $\frac{3}{6} - \frac{2}{6} = \frac{}{6}$

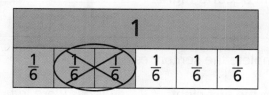

3. $\frac{8}{10} - \frac{3}{10} = \frac{}{10}$

Subtract. Use models to help.

4. $\frac{5}{8} - \frac{2}{8} =$ _____

5. $\frac{7}{12} - \frac{2}{12} =$ _____

6. $\frac{3}{4} - \frac{2}{4} =$ _____

Math Talk | MATHEMATICAL PRACTICES
Explain why the numerator changes when you subtract fractions with like denominators, but the denominator doesn't.

On Your Own

Subtract. Use models to help.

7. $\frac{2}{3} - \frac{1}{3} =$ _____

8. $\frac{7}{8} - \frac{5}{8} =$ _____

9. $\frac{4}{12} - \frac{2}{12} =$ _____

10. $\frac{2}{4} - \frac{1}{4} =$ _____

11. $\frac{9}{10} - \frac{3}{10} =$ _____

12. $1 - \frac{5}{8} =$ _____

13. $\frac{6}{8} - \frac{5}{8} =$ _____

14. $1 - \frac{1}{3} =$ _____

15. $\frac{4}{5} - \frac{2}{5} =$ _____

🗝 UNLOCK the Problem — REAL WORLD

16. Mrs. Ruiz served a pie for dessert two nights in a row. The drawings below show the pie after her family ate dessert on each night. What fraction of the pie did they eat on the second night?

First night   **Second night**

 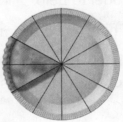

(A) $\frac{2}{12}$ (B) $\frac{5}{12}$ (C) $\frac{7}{12}$ (D) $\frac{10}{12}$

a. What do you need to know? _____

_____

b. How can you find the number of pieces eaten on the second night? _____

_____

c. Explain the steps you used to solve the problem.

_____

_____

_____

_____

_____

d. Complete the sentences.

After the first night, _____ pieces were left.

After the second night, _____ pieces were left.

So, _____ of the pie was eaten on the second night.

e. Fill in the bubble for the correct answer choice above.

17. Judi ate $\frac{7}{8}$ of a small pizza and Jack ate $\frac{2}{8}$ of a second small pizza. How much more of a pizza did Judi eat?

(A) $\frac{8}{8}$ (C) $\frac{6}{8}$

(B) 1   (D) $\frac{5}{8}$

18. Dani's bean plant grew $\frac{6}{8}$ inch the first week and $\frac{2}{8}$ inch the second week. How much less did Dani's plant grow the second week?

(A) $\frac{8}{8}$ inch   (C) $\frac{4}{8}$ inch

(B) $\frac{6}{8}$ inch   (D) $\frac{2}{8}$ inch

FOR MORE PRACTICE:
Standards Practice Book, pp. P139–P140

Name _____

Add and Subtract Fractions

Essential Question How can you add and subtract fractions with like denominators?

🗝 UNLOCK the Problem REAL WORLD

Julie is making a poster for a book report. The directions say to use $\frac{1}{5}$ of the poster to describe the setting, $\frac{2}{5}$ of the poster to describe the characters, and the rest of the poster to describe the plot. What part of the poster will she use to describe the plot?

🔑 Example Use a model.

Shade _____ to represent the part for the setting.

Shade _____ to represent the part for the characters.

1				
$\frac{1}{5}$	$\frac{1}{5}$	$\frac{1}{5}$	$\frac{1}{5}$	$\frac{1}{5}$

* Write an equation for the part of the poster used for

  the setting and characters. _____

* What does the part of the model that is not shaded represent?

  _____

* Write an equation for the part of the poster she will use for the plot.

  _____

So, Julie will use _____ of the poster to describe the plot.

Math Talk MATHEMATICAL PRACTICES
Why should Julie divide her poster into 5 equal parts instead of 3 equal parts? **Explain.**

1. **What's the Error?** Luke says $\frac{1}{5} + \frac{2}{5} = \frac{3}{10}$. **Describe** his error.

_____

_____

Common Denominators Fractions with common denominators represent wholes divided into the same number of equal-size parts. To add or subtract fractions with the same denominator, you can add or subtract the number of parts given in the numerators.

Example Complete each equation.

Words	Fractions
1 fourth-size part + 2 fourth-size parts = _____ fourth-size parts	$\frac{1}{4} + \frac{2}{4} = \frac{}{4}$
3 sixth-size parts + 2 sixth-size parts = _____	$\frac{3}{6} + \frac{2}{6} = \frac{}{}$
7 tenth-size parts − 4 tenth-size parts = _____	$\frac{}{} - \frac{}{} = \frac{}{}$

2. Sense or Nonsense? Brian says that when you add or subtract fractions with the same denominator, you can add or subtract the numerators and keep the same denominator. Is Brian correct? **Explain.**

_____

_____

_____

_____

_____

Share and Show

Math Talk MATHEMATICAL PRACTICES
Explain why $\frac{11}{12} - \frac{5}{6} \neq \frac{6}{6}$.

1. 9 twelfth-size parts − 5 twelfth-size parts = _____

$\frac{9}{12} - \frac{5}{12} =$ _____

Find the sum or difference.

2. $\frac{3}{12} + \frac{8}{12} =$ _____

3. $\frac{1}{3} + \frac{1}{3} =$ _____

4. $\frac{3}{4} - \frac{1}{4} =$ _____

⌾ 5. $\frac{2}{6} + \frac{2}{6} =$ _____

6. $\frac{3}{8} + \frac{1}{8} =$ _____

⌾ 7. $\frac{6}{10} - \frac{2}{10} =$ _____

© Houghton Mifflin Harcourt Publishing Company

Name _____

On Your Own ·

Find the sum or difference.

8. $\frac{1}{2} + \frac{1}{2} =$ _____

9. $\frac{5}{6} - \frac{4}{6} =$ _____

10. $\frac{4}{5} - \frac{2}{5} =$ _____

11. $\frac{1}{10} + \frac{3}{10} =$ _____

12. $\frac{5}{12} - \frac{1}{12} =$ _____

13. $\frac{3}{8} + \frac{2}{8} =$ _____

Practice: Copy and Solve Find the sum or difference.

14. $\frac{1}{4} + \frac{1}{4} =$ _____

15. $\frac{9}{10} - \frac{5}{10} =$ _____

16. $\frac{1}{12} + \frac{7}{12} =$ _____

 Algebra Find the unknown fraction.

17. $\frac{2}{8} +$ _____ $= \frac{6}{8}$

18. $1 -$ _____ $= \frac{3}{4}$

19. $\frac{4}{5} -$ _____ $= \frac{1}{5}$

20. **Write Math** ▸ A city worker is painting a stripe down the center of Main Street. Main Street is $\frac{8}{10}$ mile long. The worker painted $\frac{4}{10}$ mile of the street. **Explain** how to find what part of a mile is left to paint.

_____

_____

_____

21. **H.O.T.** The length of a rope was $\frac{6}{8}$ yard. Jeff cut the rope into 3 pieces. Each piece is a different length measured in eighths of a yard. What is the length of each piece of rope?

_____

22. **Test Prep** Otis has 1 cup of granola. He added $\frac{3}{8}$ cup to a bowl of yogurt. How much granola is left?

(A) $\frac{2}{8}$ cup

(C) $\frac{5}{8}$ cup

(B) $\frac{3}{8}$ cup

(D) $\frac{8}{8}$ cup

Problem Solving REAL WORLD

Sense or Nonsense?

23. Harry says that $\frac{1}{4} + \frac{1}{8} = \frac{2}{8}$. Jane says $\frac{1}{4} + \frac{1}{8} = \frac{3}{8}$.
Whose answer makes sense? Whose answer is
nonsense? **Explain** your reasoning. Draw a
model to help.

	Harry	
○	$\frac{1}{4} + \frac{1}{8} = \frac{2}{8}$	

	Jane	
○	$\frac{1}{4} + \frac{1}{8} = \frac{3}{8}$	

Model

Harry

Jane

FOR MORE PRACTICE:
Standards Practice Book, pp. P141–P142

© Houghton Mifflin Harcourt Publishing Company

Mid-Chapter Checkpoint

▶ **Vocabulary**

Choose the best term from the box.

1. A _____ always has a numerator of 1. (p. 271)

2. A number that names part of a whole is a

   _____ .

▶ **Check Concepts**

Write the fraction as a sum of unit fractions.

3. $\dfrac{3}{10} =$ _____

4. $\dfrac{6}{6} =$ _____

Use the model to write an equation.

5.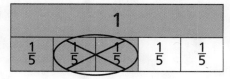

_____

6.

_____

Use the model to solve the equation.

7. $\dfrac{3}{8} + \dfrac{2}{8} =$ _____

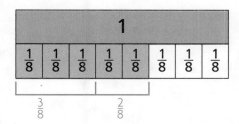

8. $\dfrac{4}{10} + \dfrac{5}{10} =$ _____

Find the sum or difference.

9. $\dfrac{9}{12} - \dfrac{7}{12} =$ _____

10. $\dfrac{2}{3} + \dfrac{1}{3} =$ _____

11. $\dfrac{1}{5} + \dfrac{3}{5} =$ _____

12. $\dfrac{2}{6} + \dfrac{2}{6} =$ _____

13. $\dfrac{4}{4} - \dfrac{2}{4} =$ _____

14. $\dfrac{7}{8} - \dfrac{4}{8} =$ _____

Fill in the bubble completely to show your answer.

15. Tyrone mixed $\frac{7}{12}$ quart of red paint with $\frac{1}{12}$ quart of yellow paint. How much paint does Tyrone have in the mixture?

Ⓐ $\frac{8}{24}$ quart

Ⓑ $\frac{6}{12}$ quart

Ⓒ $\frac{8}{12}$ quart

Ⓓ $\frac{12}{12}$ quart

16. Jorge lives $\frac{6}{8}$ mile from school and $\frac{2}{8}$ mile from a ballpark. How much farther does Jorge live from school than from the ballpark?

Ⓐ $\frac{4}{16}$ mile

Ⓑ $\frac{4}{8}$ mile

Ⓒ $\frac{8}{8}$ mile

Ⓓ 8 miles

17. Su Ling started an art project with 1 yard of felt. She used $\frac{5}{6}$ yard. How much felt does Su Ling have left?

Ⓐ $\frac{1}{6}$ yard

Ⓑ $\frac{4}{6}$ yard

Ⓒ $\frac{5}{6}$ yard

Ⓓ $\frac{6}{6}$ yard

18. Eloise hung artwork on $\frac{2}{5}$ of a bulletin board. She hung math papers on $\frac{1}{5}$ of the same bulletin board. What part of the bulletin board has artwork or math papers?

Ⓐ $\frac{1}{10}$

Ⓑ $\frac{1}{5}$

Ⓒ $\frac{3}{10}$

Ⓓ $\frac{3}{5}$

Rename Fractions and Mixed Numbers

Essential Question How can you rename mixed numbers as fractions greater than 1 and rename fractions greater than 1 as mixed numbers?

UNLOCK the Problem REAL WORLD

Mr. Fox has $2\frac{3}{6}$ loaves of corn bread. Each loaf was cut into $\frac{1}{6}$-size pieces. If he has 14 people over for dinner, is there enough bread for each person to have 1 piece?

A **mixed number** is a number represented by a whole number and a fraction. You can write a mixed number as a fraction.

To find how many $\frac{1}{6}$-size pieces are in $2\frac{3}{6}$, write $2\frac{3}{6}$ as a fraction.

- What is the size of 1 piece of bread relative to the whole?

_____

- How much bread does Mr. Fox need for 14 people?

_____

Example Write a mixed number as a fraction.

THINK

STEP 1 Model $2\frac{3}{6}$.

MODEL AND RECORD

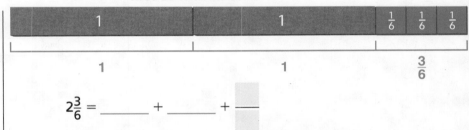

$$2\frac{3}{6} = \underline{\hspace{1cm}} + \underline{\hspace{1cm}} + \underline{\hspace{1cm}}$$

STEP 2 Find how many $\frac{1}{6}$-size pieces are in each whole. Model $2\frac{3}{6}$ using only $\frac{1}{6}$-size pieces.

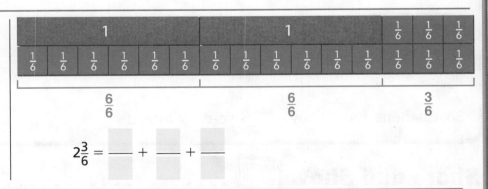

$$2\frac{3}{6} = \underline{\hspace{1cm}} + \underline{\hspace{1cm}} + \underline{\hspace{1cm}}$$

STEP 3 Find the total number of $\frac{1}{6}$-size pieces in $2\frac{3}{6}$.

Think: Find $\frac{6}{6} + \frac{6}{6} + \frac{3}{6}$.

$$2\frac{3}{6} = \underline{\hspace{1cm}}$$

There are _____ sixth-size pieces in $2\frac{3}{6}$.

So, there is enough bread for 14 people to each have 1 piece.

Math Talk MATHEMATICAL PRACTICES
Explain how to write $1\frac{1}{4}$ as a fraction without using a model.

🔒 Example   Write a fraction greater than 1 as a mixed number.

To weave a bracelet, Charlene needs 7 pieces of brown thread. Each piece of thread must be $\frac{1}{3}$ yard long. How much thread should she buy to weave the bracelet?

Write $\frac{7}{3}$ as a mixed number.

THINK   **MODEL AND RECORD**

STEP 1 Model $\frac{7}{3}$.

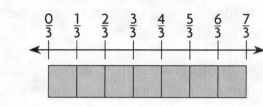

$$\frac{7}{3} = \underline{\quad} + \underline{\quad} + \underline{\quad} + \underline{\quad} + \underline{\quad} + \underline{\quad} + \underline{\quad}$$

STEP 2 Find how many wholes are in $\frac{7}{3}$, and how many thirds are left over.

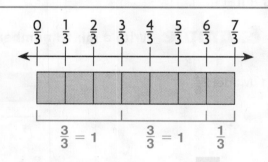

$$\frac{3}{3} = 1 \qquad \frac{3}{3} = 1 \qquad \frac{1}{3}$$

$$\frac{7}{3} = \underline{\qquad} + \underline{\qquad} + \underline{\quad}$$

STEP 3 Write $\frac{7}{3}$ as a mixed number.

$$\frac{7}{3} = \boxed{}\,\frac{\boxed{}}{\boxed{}}$$

So, Charlene should buy _____ yards of thread.

Share and Show ·

Write the unknown numbers. Write mixed numbers above the number line and fractions greater than one below the number line.

1.

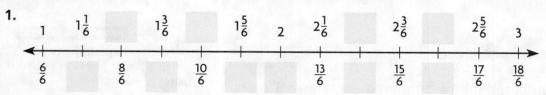

Name _____

Write the mixed number as a fraction.

2. $1\frac{1}{8}$

_____

3. $1\frac{3}{5}$

_____

✓ **4.** $1\frac{2}{3}$

_____

Write the fraction as a mixed number.

5. $\frac{11}{4}$

_____

6. $\frac{6}{5}$

_____

✓ **7.** $\frac{13}{10}$

_____

MATHEMATICAL PRACTICES

Math Talk Describe how you can compare $1\frac{3}{5}$ and $\frac{7}{5}$.

On Your Own ....

Write the mixed number as a fraction.

8. $2\frac{7}{10}$

_____

9. $3\frac{2}{3}$

_____

10. $4\frac{2}{5}$

_____

Write the fraction as a mixed number.

11. $\frac{9}{5}$

_____

12. $\frac{11}{10}$

_____

13. $\frac{12}{2}$

_____

 Algebra Find the unknown numbers.

14. $\frac{13}{7} = 1\frac{\blacksquare}{7}$

_____

15. $\blacksquare\frac{5}{6} = \frac{23}{6}$

_____

16. $\frac{57}{11} = \blacksquare\frac{\blacksquare}{11}$

_____

Problem Solving 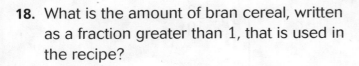 REAL WORLD

Use the recipe to solve 17–19.

Energy Squares
$1\frac{1}{3}$ cup honey
$1\frac{1}{2}$ cups peanut butter
1 cup dry milk
$3\frac{1}{4}$ cups bran cereal

17. Cal is making energy squares. How many $\frac{1}{2}$ cups of peanut butter are used in the recipe?

_____

18. What is the amount of bran cereal, written as a fraction greater than 1, that is used in the recipe?

_____

SHOW YOUR WORK

19. **H.O.T.** Suppose Cal wants to make 2 times as many energy squares as the recipe makes. How many cups of bran cereal should he use? Write your answer as a mixed number and as a fraction greater than 1 in simplest form.

_____

20. Cal added $2\frac{3}{8}$ cups of raisins. Write this mixed number as a fraction greater than 1 in simplest form.

_____

21. **H.O.T.** Jenn is preparing brown rice. She needs $1\frac{1}{2}$ cups of brown rice and 2 cups of water. Jenn has only a $\frac{1}{8}$-cup measuring cup. How many $\frac{1}{8}$ cups each of rice and water will Jenn use to prepare the rice?

_____

22. **Test Prep** Which fraction greater than 1 can you write for $4\frac{5}{9}$?

Ⓐ $\frac{9}{9}$ Ⓒ $\frac{41}{9}$

Ⓑ $\frac{18}{9}$ Ⓓ $\frac{45}{9}$

Add and Subtract Mixed Numbers

Essential Question How can you add and subtract mixed numbers with like denominators?

🔑 UNLOCK the Problem REAL WORLD

After a party, there were $1\frac{4}{6}$ quesadillas left on one tray and $2\frac{3}{6}$ quesadillas left on another tray. How much of the quesadillas were left?

- What operation will you use?

  _____

- Is the sum of the fractional parts of the mixed numbers greater than 1?

  _____

🔑 Example Add mixed numbers.

THINK	MODEL	RECORD
STEP 1 Add the fractional parts of the mixed numbers.	**Think:** Shade to model $\frac{4}{6} + \frac{3}{6}$. 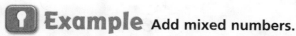	$\begin{array}{r} 1\frac{4}{6} \\ + 2\frac{3}{6} \\ \hline \end{array}$
STEP 2 Add the whole-number parts of the mixed numbers.	**Think:** Shade to model $1 + 2$.	$\begin{array}{r} 1\frac{4}{6} \\ + 2\frac{3}{6} \\ \hline \frac{7}{6} \end{array}$
STEP 3 Rename the sum.	**Think:** $\frac{7}{6}$ is greater than 1. Group the wholes together to rename the sum. The model shows a total of ___ wholes and ___ left over.	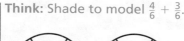 $3\frac{7}{6} = 3 + \frac{6}{6} + \underline{\quad}$ 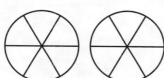 $= 3 + 1 + \underline{\quad} = \underline{\quad}$

So, _____ quesadillas were left.

Math Talk MATHEMATICAL PRACTICES
When modeling sums such as $\frac{4}{6}$ and $\frac{3}{6}$, why is it helpful to combine parts into wholes when possible? **Explain.**

🔒 Example Subtract mixed numbers.

Alejandro had $3\frac{4}{6}$ quesadillas. His family ate $2\frac{3}{6}$ of the quesadillas. How many quesadillas are left?

Find $3\frac{4}{6} - 2\frac{3}{6}$.

MODEL

Shade the model to show $3\frac{4}{6}$.

Then cross out $2\frac{3}{6}$ to model the subtraction.

The difference is _____.

So, there are _____ quesadillas left.

RECORD

Subtract the fractional parts of the mixed numbers.

Then subtract the whole-number parts of the mixed numbers.

$$
\begin{array}{r}
3\frac{4}{6} \\
- 2\frac{3}{6} \\
\hline
\end{array}
$$

Share and Show 🖊️ MATH BOARD

Write the sum as a mixed number with the fractional part less than 1.

1. $1\frac{1}{6}$ Add whole numbers.   Add fractions.

$+3\frac{3}{6}$

$+$ _____ $+$ _____

$+$ _____ $=$ _____

2. $1\frac{4}{5}$

$+7\frac{2}{5}$

✓ 3. $2\frac{1}{2}$

$+3\frac{1}{2}$

Name _____

Find the difference.

4. $3\dfrac{7}{12}$
   $-2\dfrac{5}{12}$

5. $4\dfrac{2}{3}$
   $-3\dfrac{1}{3}$

6. $6\dfrac{9}{10}$
   $-3\dfrac{7}{10}$

Math Talk MATHEMATICAL PRACTICES
Explain how adding and subtracting mixed numbers is different from adding and subtracting fractions.

On Your Own

Write the sum as a mixed number with the fractional part less than 1.

7. $7\dfrac{4}{6}$
   $+4\dfrac{3}{6}$

8. $8\dfrac{1}{3}$
   $+3\dfrac{2}{3}$

9. $5\dfrac{4}{8}$
   $+3\dfrac{5}{8}$

10. $3\dfrac{5}{12}$
    $+4\dfrac{2}{12}$

Find the difference.

11. $5\dfrac{7}{8}$
    $-2\dfrac{3}{8}$

12. $5\dfrac{7}{12}$
    $-4\dfrac{1}{12}$

13. $3\dfrac{5}{10}$
    $-1\dfrac{3}{10}$

14. $7\dfrac{3}{4}$
    $-2\dfrac{2}{4}$

Practice: Copy and Solve Find the sum or difference.

15. $1\dfrac{3}{8} + 2\dfrac{7}{8}$

16. $6\dfrac{5}{8} - 4$

17. $9\dfrac{1}{2} + 8\dfrac{1}{2}$

18. $6\dfrac{3}{5} + 4\dfrac{3}{5}$

19. $8\dfrac{7}{10} - \dfrac{4}{10}$

20. $7\dfrac{3}{5} - 6\dfrac{3}{5}$

© Houghton Mifflin Harcourt Publishing Company

Problem Solving REAL WORLD

Solve. Write your answer as a mixed number.

SHOW YOUR WORK

21. The driving distance from Alex's house to the museum is $6\frac{7}{10}$ miles. What is the round-trip distance?

_____

22. The driving distance from the sports arena to Kristina's house is $10\frac{9}{10}$ miles. The distance from the sports arena to Luke's house is $2\frac{7}{10}$ miles. How much greater is the driving distance between the sports arena and Kristina's house than between the sports arena and Luke's house?

_____

23. Benji biked from his house to the nature preserve, a distance of $23\frac{4}{5}$ miles. Jade biked from her house to the lake, a distance of $12\frac{2}{5}$ miles. How many fewer miles did Jade bike than Benji?

_____

24. **H.O.T.** During the Samson family trip, they drove from home to a ski lodge, a distance of $55\frac{4}{5}$ miles, and then drove an additional $12\frac{4}{5}$ miles to visit friends. If the family drove the same route back home, what was the distance traveled during their trip?

_____

25. **Test Prep** Jeff used $4\frac{7}{8}$ cups of orange juice and $3\frac{1}{8}$ cups of pineapple juice to make tropical punch. How much more orange juice than pineapple juice did Jeff use?

Ⓐ $\frac{3}{4}$ cup

Ⓑ $1\frac{3}{4}$ cups

Ⓒ $1\frac{7}{8}$ cups

Ⓓ 8 cups

FOR MORE PRACTICE:
Standards Practice Book, pp. P145–P146

Name _____

Subtraction with Renaming

Essential Question How can you rename a mixed number to help you subtract?

 UNLOCK the Problem REAL WORLD

Bruce, Chandler, and Chase go bike riding on weekends. On one weekend, Chase rode his bike for 3 hours, Chandler rode her bike for $2\frac{1}{4}$ hours, and Bruce rode his bike for $1\frac{3}{4}$ hours. How much longer did Chandler ride her bike than Bruce did?

- Which operation will you use?

  _____

- In the problem, circle the numbers that you need to use to find a solution.

🔒 **Use a model. Find $2\frac{1}{4} - 1\frac{3}{4}$.**

Shade the model to show how long Chandler rode her bike.

Then shade the model to show how long Bruce rode his bike.

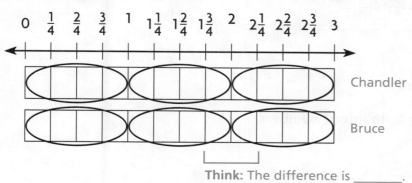

Think: The difference is _____.

So, Chandler rode her bike _____ hour longer than Bruce did.

---

1. If you have 1 fourth-size part, can you take away 3 fourth-size parts? **Explain.**

_____

2. If you have 1 whole and 1 fourth-size part, can you take away 3 fourth-size parts? **Explain.**

_____

_____

Math Talk MATHEMATICAL PRACTICES

Explain how you can find how much longer Chase rode his bike than Chandler did.

🔑 One Way **Rename the first mixed number.**

Find the difference. $5\frac{1}{8} - 3\frac{3}{8}$

STEP 1

Rename $5\frac{1}{8}$ as a mixed number with a fraction greater than 1.

Think:

$$5\frac{1}{8} = 4 + 1 + \frac{1}{8}$$

$$= 4 + \frac{}{8} + \frac{1}{8}$$

$$= $$

Math Talk MATHEMATICAL PRACTICES
Explain why you need to rename $5\frac{1}{8}$.

STEP 2

Subtract the mixed numbers.

$$5\frac{1}{8} = $$

$$-3\frac{3}{8} = -3\frac{3}{8}$$

🔑 Another Way **Rename both mixed numbers.**

Find the difference. $3\frac{4}{12} - 1\frac{6}{12}$

STEP 1

Rename both mixed numbers as fractions greater than 1.

$$3\frac{4}{12} = \frac{}{12} \qquad 1\frac{6}{12} = \frac{}{12}$$

STEP 2

Subtract the fractions greater than 1.

$$\frac{}{12}$$

$$-\frac{}{12}$$

- **Explain** how you could rename 5 to subtract $3\frac{1}{4}$.

_____

_____

Name _____

Share and Show

1. Rename both mixed numbers as fractions. Find the difference.

$$3\frac{3}{6} = \frac{}{6}$$

$$-1\frac{4}{6} = -\frac{}{6}$$

Find the difference.

2. $1\frac{1}{3}$

 $-\frac{2}{3}$

3. $4\frac{7}{10}$

 $-1\frac{9}{10}$

4. $3\frac{5}{12}$

 $-\frac{8}{12}$

Math Talk MATHEMATICAL PRACTICES

Describe how you would model $\frac{13}{6} - \frac{8}{6}$.

On Your Own

Find the difference.

5. $8\frac{1}{10}$

 $-2\frac{9}{10}$

6. 2

 $-1\frac{1}{4}$

7. $4\frac{1}{5}$

 $-3\frac{2}{5}$

Practice: Copy and Solve **Find the difference.**

8. $4\frac{1}{6} - 2\frac{5}{6}$

9. $6\frac{9}{12} - 3\frac{10}{12}$

10. $3\frac{3}{10} - \frac{7}{10}$

11. $4 - 2\frac{3}{5}$

12. $5\frac{1}{4} - 2\frac{3}{4}$

13. $3\frac{9}{12} - 1\frac{11}{12}$

14. $7\frac{3}{10} - 4\frac{7}{10}$

15. $2\frac{3}{8} - 1\frac{5}{8}$

Problem Solving REAL WORLD

Rename the fractions to solve.

Many instruments are coiled or curved so that they are easier for the musician to play, but they would be quite long if straightened out completely.

16. Trumpets and cornets are brass instruments. Fully stretched out, the length of a trumpet is $5\frac{1}{4}$ feet and the length of a cornet is $4\frac{2}{4}$ feet. The trumpet is how much longer than the cornet?

_____

17. Bassoons and double bassoons are woodwind instruments. Fully stretched out, the length of a bassoon is $9\frac{2}{12}$ feet and the length of a double bassoon is $18\frac{1}{12}$ feet. The double bassoon is how much longer than the bassoon?

_____

SHOW YOUR WORK

18. Tubas, trombones, and French horns are brass instruments. Fully stretched out, the length of a tuba is 18 feet, the length of a trombone is $9\frac{11}{12}$ feet, and the length of a French horn is $17\frac{1}{12}$ feet. The tuba is how much longer than the French horn? The French horn is how much longer than the trombone?

_____

19. H.O.T. The pitch of a musical instrument is related to its length. In general, the greater the length of a musical instrument, the lower its pitch. Order the brass instruments identified on this page from lowest pitch to the highest pitch.

_____

20. **Test Prep** Sam ran $2\frac{2}{5}$ miles on Saturday and $1\frac{4}{5}$ miles on Monday. How many more miles did Sam run on Saturday than on Monday?

Ⓐ $\frac{3}{5}$ mile   Ⓒ $1\frac{2}{5}$ miles

Ⓑ $\frac{4}{5}$ mile   Ⓓ $1\frac{4}{5}$ miles

Name _____

Fractions and Properties of Addition

Essential Question How can you add fractions with like denominators using the properties of addition?

CONNECT The Associative and Commutative Properties of Addition can help you group and order addends to find sums mentally. You can use mental math to combine fractions that have a sum of 1.

- The Commutative Property of Addition states that when the order of two addends is changed, the sum is the same. For example, $4 + 5 = 5 + 4$.

- The Associative Property of Addition states that when the grouping of addends is changed, the sum is the same. For example, $(5 + 8) + 4 = 5 + (8 + 4)$.

🔑 UNLOCK the Problem REAL WORLD

The map shows four lighthouses in the Florida Keys and their distances apart in miles. The Dry Tortugas Lighthouse is the farthest west, and the Alligator Reef Lighthouse is the farthest east.

What is the distance from the Dry Tortugas Lighthouse to the Alligator Reef Lighthouse, traveling between the four lighthouses?

Gulf of Mexico

$70\frac{5}{10}$ $43\frac{6}{10}$ $34\frac{5}{10}$

Dry Tortugas Lighthouse

Key West Lighthouse

Sombrero Key Lighthouse

Alligator Reef Lighthouse

🔓 **Use the properties to order and group.**

Add. $70\frac{5}{10} + 43\frac{6}{10} + 34\frac{5}{10}$

$70\frac{5}{10} + 43\frac{6}{10} + 34\frac{5}{10} =$ _____ + _____ + _____

$= ($_____ + _____$) +$ _____

$= ($_____$) +$ _____

$=$ _____

Use the Commutative Property to order the addends so that the fractions with a sum of 1 are together.

Use the Associative Property to group the addends that you can add mentally.

Add the grouped numbers, and then add the other mixed number.

Write the sum.

So, the distance from the Dry Tortugas Lighthouse to the Alligator Reef Lighthouse, traveling between the four lighthouses, is

_____ miles.

© Houghton Mifflin Harcourt Publishing Company

Try This! Use the properties and mental math to solve. Show each step, and name the property used.

$$1\tfrac{1}{3} + (2 + 3\tfrac{2}{3})$$

Share and Show

1. Complete. Name the property used.

$$\left(3\tfrac{4}{10} + 5\tfrac{2}{10}\right) + \tfrac{6}{10} = \left(5\tfrac{2}{10} + 3\tfrac{4}{10}\right) + \underline{\qquad} \quad \underline{\hspace{4cm}}$$

$$= 5\tfrac{2}{10} + \left(3\tfrac{4}{10} + \underline{\qquad}\right) \quad \underline{\hspace{4cm}}$$

$$= 5\tfrac{2}{10} + \underline{\qquad}$$

$$= \underline{\qquad}$$

> **MATHEMATICAL PRACTICES**
>
> **Math Talk** Describe how you could use the properties to find the sum $1\tfrac{1}{3} + 2\tfrac{5}{8} + 1\tfrac{2}{3}$.

Use the properties and mental math to find the sum.

2. $\left(2\tfrac{7}{8} + 3\tfrac{2}{8}\right) + 1\tfrac{1}{8}$

3. $1\tfrac{2}{5} + \left(1 + \tfrac{3}{5}\right)$

4. $5\tfrac{3}{6} + \left(5\tfrac{5}{6} + 4\tfrac{3}{6}\right)$

⏣ 5. $\left(1\tfrac{1}{4} + 1\tfrac{1}{4}\right) + 2\tfrac{3}{4}$

6. $\left(12\tfrac{4}{9} + 1\tfrac{2}{9}\right) + 3\tfrac{5}{9}$

⏣ 7. $\tfrac{3}{12} + \left(1\tfrac{8}{12} + \tfrac{9}{12}\right)$

Name _____

On Your Own...

Use the properties and mental math to find the sum.

8. $\left(45\frac{1}{3} + 6\frac{1}{3}\right) + 38\frac{2}{3}$

9. $\frac{1}{2} + \left(103\frac{1}{2} + 12\right)$

10. $\left(3\frac{5}{10} + 10\right) + 11\frac{5}{10}$

11. $1\frac{4}{10} + \left(37\frac{3}{10} + \frac{6}{10}\right)$

12. $\left(\frac{3}{12} + 10\frac{5}{12}\right) + \frac{9}{12}$

13. $5\frac{7}{8} + \left(6\frac{3}{8} + \frac{1}{8}\right)$

Use the expressions in the box for 14–15.

14. Which property of addition would be best to solve Expression A?

_____

15. **H.O.T.** Which two expressions have the same value?

_____

16. **Test Prep** Shari represented the perimeter of a triangle she measured using the expression $1\frac{1}{2} + \left(2 + 2\frac{1}{2}\right)$. How could Shari rewrite the expression using both the Associative and Commutative Properties?

Ⓐ $3 + 2\frac{1}{2}$

Ⓑ $1\frac{1}{2} + \left(2\frac{1}{2} + 2\right)$

Ⓒ $\left(1\frac{1}{2} + 2\right) + 2\frac{1}{2}$

Ⓓ $\left(1\frac{1}{2} + 2\frac{1}{2}\right) + 2$

A $\frac{1}{8} + \left(\frac{7}{8} + \frac{4}{8}\right)$
B $\frac{1}{2} + 2$
C $\frac{3}{7} + \left(\frac{1}{2} + \frac{4}{7}\right)$
D $\frac{1}{3} + \frac{4}{3} + \frac{2}{3}$

Problem Solving REAL WORLD

Pose a Problem

17. Costumes are being made for the high school musical. The table at the right shows the amount of fabric needed for the costumes of the male and female leads. Alice uses the expression $7\frac{3}{8} + 1\frac{5}{8} + 2\frac{4}{8}$ to find the total amount of fabric needed for the costume of the female lead.

To find the value of the expression using mental math, Alice used the properties of addition.

$$7\frac{3}{8} + 1\frac{5}{8} + 2\frac{4}{8} = \left(7\frac{3}{8} + 1\frac{5}{8}\right) + 2\frac{4}{8}$$

Alice added $7 + 1$ and was able to quickly add $\frac{3}{8}$ and $\frac{5}{8}$ to the sum of 8 to get 9. She added $2\frac{4}{8}$ to 9, so her answer was $11\frac{4}{8}$.

So, the amount of fabric needed for the costume of the female lead actor is $11\frac{4}{8}$ yards.

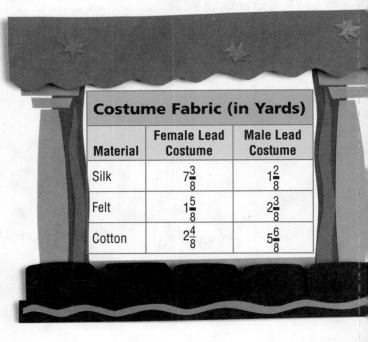

Costume Fabric (in Yards)

Material	Female Lead Costume	Male Lead Costume
Silk	$7\frac{3}{8}$	$1\frac{2}{8}$
Felt	$1\frac{5}{8}$	$2\frac{3}{8}$
Cotton	$2\frac{4}{8}$	$5\frac{6}{8}$

Write a new problem using the information for the costume for the male lead actor.

Pose a Problem	Solve your problem. Check your solution.

- **Explain** how using the properties of addition makes both problems easier to solve.

 FOR MORE PRACTICE:
Standards Practice Book, pp. P149–P150

Problem Solving • Multistep Fraction Problems

Essential Question How can you use the strategy *act it out* to solve multistep problems with fractions?

🔑 UNLOCK the Problem · REAL WORLD

A gift shop sells walnuts in $\frac{3}{4}$-pound bags. Ann will buy some bags of walnuts and repackage them into 1-pound bags. What is the least number of $\frac{3}{4}$-pound bags Ann could buy, if she wants to fill each 1-pound bag, without leftovers?

Read the Problem

What do I need to find?

I need to find how many

_____ bags of walnuts Ann needs to make 1-pound bags of walnuts, without leftovers.

What information do I need to use?

The bags she will buy contain

_____ pound of walnuts. She will repackage the walnuts into

_____ -pound bags.

How will I use the information?

I can use fraction circles to

_____ the problem.

Solve the Problem

Describe how to act it out. Use fraction circles.

One $\frac{3}{4}$-pound bag   Not enough for a 1-pound bag

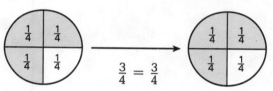

$$\frac{3}{4} = \frac{3}{4}$$

Two $\frac{3}{4}$-pound bags   One 1-pound bag with $\frac{2}{4}$ pound left over

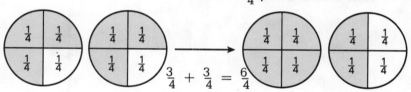

$$\frac{3}{4} + \frac{3}{4} = \frac{6}{4}$$

Three $\frac{3}{4}$-pound bags have $\frac{3}{4} + \frac{3}{4} + \frac{3}{4} = \frac{\quad}{4}$ pounds of

walnuts. This makes _____ 1-pound bags with _____ pound left over.

Four $\frac{3}{4}$-pound bags have $\frac{3}{4} + \frac{3}{4} + \frac{3}{4} + \frac{3}{4} = \frac{\quad}{4}$-pounds of walnuts.

This makes _____ 1-pound bags with _____ left over.

So, Ann could buy _____ $\frac{3}{4}$-pound bags of walnuts.

🔒 Try Another Problem

At the end of dinner, a restaurant had several dishes of quiche, each with 2 sixth-size pieces of quiche. The chef was able to combine these pieces to make 2 whole quiches, with no leftovers. How many dishes did the chef combine?

Read the Problem	Solve the Problem
What do I need to find?	**Describe how to act it out.**
What information do I need to use?	
How will I use the information?	

So, the chef combined _____ dishes each with $\frac{2}{6}$ quiche.

Name _____

Share and Show ....

♦ UNLOCK the Problem Tips
√ Underline the question.
√ Circle the important facts.
√ Cross out unneeded information.

1. Last week, Sia ran $1\frac{1}{4}$ miles each day for 5 days and then took 2 days off. Did she run at least 6 miles last week?

   First, model the problem. Describe your model.

   _____

   _____

   _____

   _____

   Then, regroup the parts in the model to find the number of whole miles Sia ran.

   Sia ran _____ whole miles and _____ mile.

   Finally, compare the total number of miles she ran to 6 miles.

   $6\frac{1}{4}$ miles ◯ 6 miles

   So, Sia _____ run at least 6 miles last week.

SHOW YOUR WORK

2. **What if** Sia ran only ran $\frac{3}{4}$ mile each day. Would she have run at least 6 miles last week? **Explain.**

   _____

   _____

3. A quarter is $\frac{1}{4}$ dollar. Noah has 20 quarters. How much money does he have? **Explain.**

   _____

   _____

   _____

4. **H.O.T.** How many $\frac{2}{5}$ parts are in 2 wholes?

   _____

On Your Own........

5. A company shipped 15,325 boxes of apples and 12,980 boxes of oranges. How many more boxes of apples than oranges did the company ship?

_____

6. A fair sold a total of 3,300 tickets on Friday and Saturday. It sold 100 more on Friday than on Saturday. How many tickets did the fair sell on Friday?

_____

7. **Write Math** ➤ Emma walked $\frac{1}{4}$ mile on Monday, $\frac{2}{4}$ mile on Tuesday, and $\frac{3}{4}$ mile on Wednesday. If the pattern continues, how many miles will she walk on Friday? **Explain** how you found the number of miles.

_____

_____

_____

_____

_____

8. **H.O.T.** Jared painted a mug $\frac{5}{12}$ red and $\frac{4}{12}$ blue. What part of the mug is **not** red or blue?

_____

9. **Test Prep** Sophia spent $\frac{3}{4}$ hour reading about the Warwick Applefest. Then she spent $\frac{1}{4}$ hour writing a paragraph about the festival. How long did Sophia spend reading and writing about the festival?

Ⓐ $\frac{1}{4}$ hour   Ⓒ 1 hour

Ⓑ $\frac{2}{4}$ hour   Ⓓ $1\frac{1}{4}$ hours

Choose a STRATEGY

Act It Out
Draw a Diagram
Find a Pattern
Make a Table or List
Solve a Simpler Problem

SHOW YOUR WORK

© Houghton Mifflin Harcourt Publishing Company

FOR MORE PRACTICE:
Standards Practice Book, pp. P151–P152

Name _____

Chapter Review/Test

▶ Vocabulary

Choose the best term from the box.

Vocabulary
mixed number
simplest form
unit fraction

1. A number represented by a whole number and a fraction is a

   _____ . (p. 289)

2. A fraction that always has a numerator of 1 is a

   _____ . (p. 271)

▶ Check Concepts

Write the fraction as a sum of unit fractions.

3. $\dfrac{4}{5} =$ _____

4. $\dfrac{5}{10} =$ _____

Write the mixed number as a fraction.

5. $1\dfrac{3}{8} =$ _____

6. $4\dfrac{2}{3} =$ _____

7. $2\dfrac{3}{5} =$ _____

Write the fraction as a mixed number.

8. $\dfrac{12}{10} =$ _____

9. $\dfrac{10}{3} =$ _____

10. $\dfrac{15}{6} =$ _____

Find the sum or difference.

11. $2\dfrac{3}{8} + 1\dfrac{6}{8} =$ _____

12. $\dfrac{9}{12} - \dfrac{2}{12} =$ _____

13. $5\dfrac{7}{10} - 4\dfrac{5}{10} =$ _____

14. $4\dfrac{1}{6} - 2\dfrac{5}{6} =$ _____

15. $3\dfrac{2}{5} - 1\dfrac{4}{5} =$ _____

16. $\dfrac{4}{12} + \dfrac{6}{12} =$ _____

Use the properties and mental math to find the sum.

17. $\left(1\dfrac{2}{5} + \dfrac{1}{5}\right) + 2\dfrac{3}{5}$

18. $2\dfrac{4}{6} + \left(2\dfrac{3}{6} + 2\dfrac{2}{6}\right)$

19. $\dfrac{3}{10} + \left(2\dfrac{4}{10} + \dfrac{7}{10}\right)$

_____

_____

_____

GO Online Assessment Options **Chapter Test**

Fill in the bubble completely to show your answer.

20. Eddie cut $2\frac{2}{4}$ feet of balsa wood for the length of a kite. He cut $\frac{3}{4}$ foot for the width of the kite. How much longer is the length of the kite than the width?

Ⓐ $1\frac{1}{4}$ feet

Ⓑ $1\frac{3}{4}$ feet

Ⓒ 2 feet

Ⓓ $3\frac{1}{4}$ feet

21. On a trip to the art museum, Lily rode the subway for $\frac{7}{10}$ mile and walked for $\frac{3}{10}$ mile. How much farther did she ride on the subway than walk?

Ⓐ $\frac{3}{10}$ mile

Ⓑ $\frac{4}{10}$ mile

Ⓒ $\frac{7}{10}$ mile

Ⓓ 1 mile

22. Pablo is training for a marathon. He ran $5\frac{4}{8}$ miles on Friday, $6\frac{5}{8}$ miles on Saturday, and $7\frac{4}{8}$ miles on Sunday. How many miles did he run on all three days ?

Ⓐ $1\frac{5}{8}$ miles

Ⓑ $12\frac{1}{8}$ miles

Ⓒ $19\frac{4}{8}$ miles

Ⓓ $19\frac{5}{8}$ miles

23. Cindy has two jars of paint.

Which fraction below represents how much paint Cindy has?

Ⓐ $\frac{1}{8}$

Ⓒ $\frac{5}{8}$

Ⓑ $\frac{4}{8}$

Ⓓ $\frac{7}{8}$

Name _____

24. Cole grew $2\frac{3}{4}$ inches last year. Kelly grew the same amount. Which fraction below represents the number of inches that Kelly grew last year?

Ⓐ $\frac{3}{4}$

Ⓑ $\frac{5}{4}$

Ⓒ $\frac{11}{4}$

Ⓓ $\frac{14}{4}$

25. Olivia's dog is 4 years old. Her cat is $1\frac{1}{2}$ years younger. How old is Olivia's cat?

Ⓐ $5\frac{1}{2}$ years old

Ⓑ $3\frac{1}{2}$ years old

Ⓒ $2\frac{1}{2}$ years old

Ⓓ $1\frac{1}{2}$ years old

26. Lisa mixed $4\frac{4}{6}$ cups of orange juice with $3\frac{1}{6}$ cups of milk to make a health shake. She drank $3\frac{3}{6}$ cups of the health shake. How much of the health shake did Lisa not drink?

Ⓐ $\frac{2}{6}$ cups

Ⓑ $4\frac{2}{6}$ cups

Ⓒ $7\frac{5}{6}$ cups

Ⓓ $11\frac{2}{6}$ cups

27. Keiko entered a contest to design a new school flag. Five twelfths of her flag has stars and $\frac{3}{12}$ has stripes. What fraction of Keiko's flag has stars and stripes?

Ⓐ $\frac{8}{12}$

Ⓑ $\frac{8}{24}$

Ⓒ $\frac{2}{12}$

Ⓓ $\frac{2}{24}$

▶ Constructed Response

28. Ela is knitting a scarf from a pattern. The pattern calls for $4\frac{2}{12}$ yards of yarn. She has only $2\frac{11}{12}$ yards of yarn. How much more yarn does Ela need to finish knitting the scarf? **Explain** how you found your answer.

_____

_____

_____

▶ Performance Task

29. Miguel's class went to the state fair. The fairground is divided into sections. Rides are in $\frac{6}{10}$ of the fairground. Games are in $\frac{2}{10}$ of the fairground. Farm exhibits are in $\frac{1}{10}$ of the fairground.

Ⓐ How much greater is the fraction of the fairground with rides than the fraction with farm exhibits? Draw a model to prove your answer is correct.

_____

Ⓑ What fraction of the fairground has games and farm exhibits? Write an equation to show your answer.

_____

Ⓒ The rest of the fairground is refreshment booths. What fraction of the fairground is refreshment booths? **Describe** the steps you follow to solve the problem.

_____

_____

_____

_____

Name _____

Multiples of Unit Fractions

Essential Question How can you write a fraction as a product of a whole number and a unit fraction?

🔑 UNLOCK the Problem REAL WORLD

At a pizza party, each pizza was cut into 6 equal slices. At the end of the party, there was $\frac{5}{6}$ of a pizza left. Roberta put each of the leftover slices in its own freezer bag. How many bags did she use? What part of a pizza did she put in each bag?

- How many slices of pizza were eaten?

  _____

- What fraction of the pizza is 1 slice?

  _____

🔑 Example Write $\frac{5}{6}$ as the product of a whole number and a unit fraction.

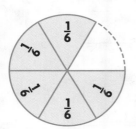

The picture shows $\frac{5}{6}$ or

_____ sixth-size parts.

Each sixth-size part of the pizza can be shown by the

unit fraction _____ .

You can use unit fractions to show $\frac{5}{6}$ in two ways.

$\frac{5}{6} = $ _____ + _____ + _____ + _____ + _____

$\frac{5}{6} = $ _____ $\times \frac{1}{6}$

The number of addends, or the multiplier, represents the number of bags used.

The unit fractions represent the part of a pizza in each bag.

So, Roberta used _____ bags. She put _____ of a pizza in each bag.

Remember

You can use multiplication to show repeated addition.

3×4 means $4 + 4 + 4$.

4×2 means $2 + 2 + 2 + 2$.

Math Talk MATHEMATICAL PRACTICES
Explain how you can write $\frac{3}{2}$ as a mixed number.

- **Explain** how you can write $\frac{3}{2}$ as the product of a whole number and a unit fraction.

  _____

  _____

  _____

Multiples The product of a number and a counting number is a multiple of the number. You have learned about multiples of whole numbers.

The products 1×4, 2×4, 3×4, and so on are multiples of 4. The numbers 4, 8, 12, and so on are multiples of 4.

You can also find multiples of unit fractions.

🔑 $1 \times \frac{1}{4}$ is $\frac{1}{4}$. **Use models to write the next four multiples of $\frac{1}{4}$. Complete the last model.**

$\frac{1}{4}$	$\frac{1}{4}$	$\frac{1}{4}$	$\frac{1}{4}$
$\frac{1}{4}$	$\frac{1}{4}$	$\frac{1}{4}$	$\frac{1}{4}$

$2 \times \frac{1}{4} = \frac{2}{4}$

$\frac{1}{4}$	$\frac{1}{4}$	$\frac{1}{4}$	$\frac{1}{4}$
$\frac{1}{4}$	$\frac{1}{4}$	$\frac{1}{4}$	$\frac{1}{4}$
$\frac{1}{4}$	$\frac{1}{4}$	$\frac{1}{4}$	$\frac{1}{4}$

$3 \times \underline{\quad} = \dfrac{\boxed{}}{4}$

$\frac{1}{4}$	$\frac{1}{4}$	$\frac{1}{4}$	$\frac{1}{4}$
$\frac{1}{4}$	$\frac{1}{4}$	$\frac{1}{4}$	$\frac{1}{4}$
$\frac{1}{4}$	$\frac{1}{4}$	$\frac{1}{4}$	$\frac{1}{4}$
$\frac{1}{4}$	$\frac{1}{4}$	$\frac{1}{4}$	$\frac{1}{4}$

$4 \times \underline{\quad} = \dfrac{\boxed{}}{4}$

$\frac{1}{4}$	$\frac{1}{4}$	$\frac{1}{4}$	$\frac{1}{4}$
$\frac{1}{4}$	$\frac{1}{4}$	$\frac{1}{4}$	$\frac{1}{4}$
$\frac{1}{4}$	$\frac{1}{4}$	$\frac{1}{4}$	$\frac{1}{4}$
$\frac{1}{4}$	$\frac{1}{4}$	$\frac{1}{4}$	$\frac{1}{4}$
$\frac{1}{4}$	$\frac{1}{4}$	$\frac{1}{4}$	$\frac{1}{4}$

$\boxed{} \times \underline{\quad} = \underline{\quad}$

Multiples of $\frac{1}{4}$ are $\frac{1}{4}$, ▨ , ▨ , ▨ , and ▨ .

🔑 **Use a number line to write multiples of $\frac{1}{5}$.**

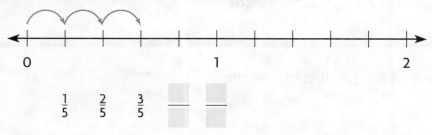

$\frac{1}{5}$ $\frac{2}{5}$ $\frac{3}{5}$ $\dfrac{\boxed{}}{\boxed{}}$ $\dfrac{\boxed{}}{\boxed{}}$

Multiples of $\frac{1}{5}$ are $\frac{1}{5}$, ▨ , ▨ , ▨ , and ▨ .

Name _____

Share and Show

1. Use the picture to complete the equations.

$\dfrac{3}{4} =$ _____ + _____ + _____

$\dfrac{3}{4} =$ _____ $\times \dfrac{1}{4}$

Write the fraction as a product of a whole number and a unit fraction.

2. $\dfrac{4}{5} =$ _____

✓ 3. $\dfrac{3}{10} =$ _____

4. $\dfrac{8}{3} =$ _____

List the next four multiples of the unit fraction.

5. $\dfrac{1}{6},$ ▢ , ▢ , ▢ , ▢

✓ 6. $\dfrac{1}{3},$ ▢ , ▢ , ▢ , ▢

Math Talk MATHEMATICAL PRACTICES Explain why $\dfrac{8}{5}$ is a multiple of $\dfrac{1}{5}$.

On Your Own

Write the fraction as a product of a whole number and a unit fraction.

7. $\dfrac{7}{8} =$ _____

8. $\dfrac{1}{12} =$ _____

9. $\dfrac{6}{5} =$ _____

10. $\dfrac{5}{6} =$ _____

11. $\dfrac{9}{4} =$ _____

12. $\dfrac{3}{100} =$ _____

List the next four multiples of the unit fraction.

13. $\dfrac{1}{10},$ ▢ , ▢ , ▢ , ▢

14. $\dfrac{1}{8},$ ▢ , ▢ , ▢ , ▢

Problem Solving REAL WORLD

15. **H.O.T.** **Write Math** Robyn uses $\dfrac{1}{2}$ cup of blueberries to make each loaf of blueberry bread. **Explain** how many loaves of blueberry bread she can make with $2\dfrac{1}{2}$ cups of blueberries.

_____

_____

16. **Test Prep** Tom makes 8 loaves of blueberry bread. He uses $\dfrac{1}{2}$ cup of blueberries in each loaf. How many cups of blueberries does he use for the 8 loaves?

(A) $\dfrac{5}{2}$ (B) $\dfrac{6}{2}$ (C) $\dfrac{7}{2}$ (D) $\dfrac{8}{2}$

Sense or Nonsense?

17. Whose statement makes sense? Whose statement is nonsense? Explain your reasoning.

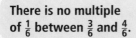

There is no multiple of $\frac{1}{6}$ between $\frac{3}{6}$ and $\frac{4}{6}$.

$\frac{4}{5}$ is a multiple of $\frac{1}{4}$.

Gavin	Abigail

- For the statement that is nonsense, write a new statement that makes sense.

FOR MORE PRACTICE:
Standards Practice Book, pp. P157–P158

Name _____

Multiples of Fractions

Essential Question How can you write a product of a whole number and a fraction as a product of a whole number and a unit fraction?

UNLOCK the Problem REAL WORLD

Jen is making 4 pans of baked ziti. For each pan, she needs $\frac{2}{3}$ cup cheese. Her measuring cup can scoop $\frac{1}{3}$ cup of cheese. How many scoops of cheese does she need for the 4 pans?

Example 1 Use a model to write the product of $4 \times \frac{2}{3}$ as the product of a whole number and a unit fraction.

$\frac{1}{3}$	$\frac{1}{3}$	$\frac{1}{3}$

Think: $\frac{2}{3}$ is 2 third-size parts.

$\frac{2}{3} = $ _____ + _____ or $2 \times$ _____ .

There are 4 pans of baked ziti. Each pan needs $\frac{2}{3}$ cup cheese.

$\frac{1}{3}$	$\frac{1}{3}$	$\frac{1}{3}$

← 1 pan: $2 \times \frac{1}{3} = \frac{2}{3}$

$\frac{1}{3}$	$\frac{1}{3}$	$\frac{1}{3}$

← 2 pans: $2 \times 2 \times \frac{1}{3} = 4 \times \frac{1}{3} = \frac{4}{3}$

$\frac{1}{3}$	$\frac{1}{3}$	$\frac{1}{3}$

← 3 pans: $3 \times 2 \times \frac{1}{3} = 6 \times \frac{1}{3} = \frac{6}{3}$

$\frac{1}{3}$	$\frac{1}{3}$	$\frac{1}{3}$

← 4 pans: $4 \times 2 \times \frac{1}{3} = 8 \times \frac{1}{3} = \frac{8}{3}$

MATHEMATICAL PRACTICES

Math Talk Explain how this model of $4 \times \frac{2}{3}$ is related to a model of 4×2.

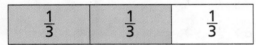

$4 \times \frac{2}{3} = 4 \times$ _____ $\times \frac{1}{3} = $ _____ $\times \frac{1}{3} = \frac{\boxed{}}{3}$

So, Jen needs _____ third-size scoops of cheese for 4 pans of ziti.

---

1. What if Jen decides to make 10 pans of ziti? **Describe** a pattern you could use to find the number of scoops of cheese she would need.

_____

_____

_____

Multiples You have learned to write multiples of unit fractions.
You can also write multiples of non-unit fractions.

🔑 **Example 2** Use a number line to write multiples of $\frac{2}{5}$.

Think: Multiply $\frac{2}{5}$ by counting numbers.

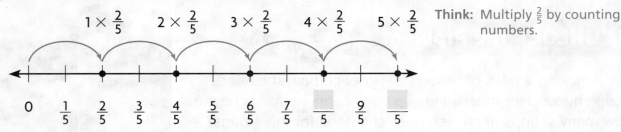

$$1 \times \frac{2}{5} \quad 2 \times \frac{2}{5} \quad 3 \times \frac{2}{5} \quad 4 \times \frac{2}{5} \quad 5 \times \frac{2}{5}$$

Multiples of $\frac{2}{5}$ are $\frac{2}{5}$, ▢ , ▢ , ▢ , and ▢ .

$3 \times \frac{2}{5} = \frac{6}{5}$. Write $\frac{6}{5}$ as a product of a whole number and a unit fraction.

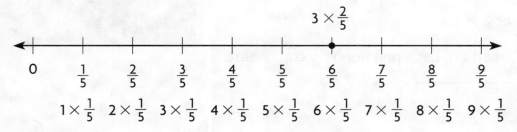

$$3 \times \frac{2}{5}$$

$$1 \times \frac{1}{5} \quad 2 \times \frac{1}{5} \quad 3 \times \frac{1}{5} \quad 4 \times \frac{1}{5} \quad 5 \times \frac{1}{5} \quad 6 \times \frac{1}{5} \quad 7 \times \frac{1}{5} \quad 8 \times \frac{1}{5} \quad 9 \times \frac{1}{5}$$

$3 \times \frac{2}{5} = \frac{6}{5} = $ _____ $\times$ _____

2. **Explain** how to use repeated addition to write the multiple of a
fraction as the product of a whole number and a unit fraction.

_____

_____

Share and Show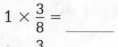

1. Write three multiples of $\frac{3}{8}$.

$1 \times \frac{3}{8} = $ _____

$2 \times \frac{3}{8} = $ _____

$3 \times \frac{3}{8} = $ _____

Multiples of $\frac{3}{8}$ are _____, _____, and _____.

$$1 \times \frac{3}{8} \quad 2 \times \frac{3}{8} \quad 3 \times \frac{3}{8}$$

320

Name _____

List the next four multiples of the fraction.

2. $\frac{3}{6}$, ____, ____, ____, ____

3. $\frac{2}{10}$, ____, ____, ____, ____

Write the product as the product of a whole number and a unit fraction.

4.
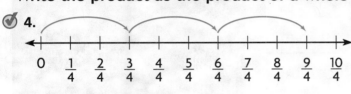

0    $\frac{1}{4}$ $\frac{2}{4}$ $\frac{3}{4}$ $\frac{4}{4}$ $\frac{5}{4}$ $\frac{6}{4}$ $\frac{7}{4}$ $\frac{8}{4}$ $\frac{9}{4}$ $\frac{10}{4}$

$3 \times \frac{3}{4} = $ _____

5.

0    $\frac{1}{6}$ $\frac{2}{6}$ $\frac{3}{6}$ $\frac{4}{6}$ $\frac{5}{6}$ $\frac{6}{6}$ $\frac{7}{6}$ $\frac{8}{6}$ $\frac{9}{6}$ $\frac{10}{6}$

$2 \times \frac{4}{6} = $ _____

MATHEMATICAL PRACTICES

Math Talk **Explain** how to write a product of a whole number and a fraction as a product of a whole number and a unit fraction.

On Your Own . . . .

List the next four multiples of the fraction.

6. $\frac{4}{5}$, ____, ____, ____, ____

7. $\frac{2}{4}$, ____, ____, ____, ____

8. $\frac{5}{12}$, ____, ____, ____, ____

9. $\frac{3}{8}$, ____, ____, ____, ____

Write the product as the product of a whole number and a unit fraction.

10.
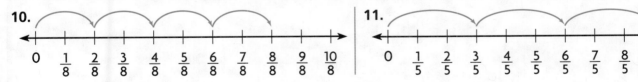

0    $\frac{1}{8}$ $\frac{2}{8}$ $\frac{3}{8}$ $\frac{4}{8}$ $\frac{5}{8}$ $\frac{6}{8}$ $\frac{7}{8}$ $\frac{8}{8}$ $\frac{9}{8}$ $\frac{10}{8}$

$4 \times \frac{2}{8} = $ _____

11.
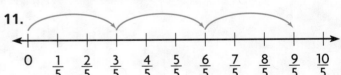

0    $\frac{1}{5}$ $\frac{2}{5}$ $\frac{3}{5}$ $\frac{4}{5}$ $\frac{5}{5}$ $\frac{6}{5}$ $\frac{7}{5}$ $\frac{8}{5}$ $\frac{9}{5}$ $\frac{10}{5}$

$3 \times \frac{3}{5} = $ _____

12. **H.O.T.** Are $\frac{6}{10}$ and $\frac{6}{30}$ multiples of $\frac{3}{10}$? **Explain.**

_____

_____

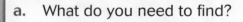

🔑 UNLOCK the Problem REAL WORLD

13. Josh is watering his plants. He gives each of 2 plants $\frac{3}{5}$ pint of water. His watering can holds $\frac{1}{5}$ pint. How many times will he fill his watering can to water both plants?

a. What do you need to find?

_____

_____

b. What information do you need to use?

_____

_____

c. How can drawing a model help you solve the problem?

_____

_____

_____

d. Show the steps you use to solve the problem.

e. Complete the sentence.

Josh will fill his watering can _____ times.

14. What is $\frac{5}{12}$ written as a product of a whole number and a unit fraction?

_____

15. **Test Prep** Which is a multiple of $\frac{5}{6}$?

Ⓐ $\frac{5}{12}$ Ⓒ $\frac{10}{5}$

Ⓑ $\frac{10}{12}$ Ⓓ $\frac{10}{6}$

Name _____

 Mid-Chapter Checkpoint

▶ **Vocabulary**

Choose the best term from the box.

1. A _____ of a number is the product of the number and a counting number. (p. 316)

2. A _____ always has a numerator of 1. (p. 314)

▶ **Concepts and Skills**

List the next four multiples of the unit fraction.

3. $\frac{1}{2}$, ▢ , ▢ , ▢ , ▢

4. $\frac{1}{5}$, ▢ , ▢ , ▢ , ▢

Write the fraction as a product of a whole number and a unit fraction.

5. $\frac{4}{10} =$ _____

6. $\frac{8}{12} =$ _____

7. $\frac{3}{4} =$ _____

List the next four multiples of the fraction.

8. $\frac{2}{5}$, ▢ , ▢ , ▢ , ▢

9. $\frac{5}{6}$, ▢ , ▢ , ▢ , ▢

Write the product as the product of a whole number and a unit fraction.

10.

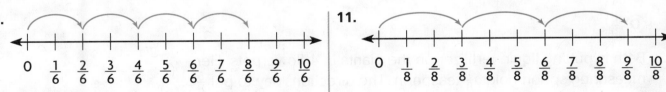

$4 \times \frac{2}{6} =$ _____

11.

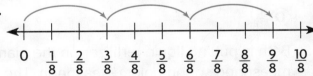

$3 \times \frac{3}{8} =$ _____

Fill in the bubble completely to show your answer.

12. Pedro cut a sheet of poster board into 10 equal parts. His brother used some of the poster board and now $\frac{8}{10}$ is left. Pedro wants to make a sign from each remaining part of the poster board. How many signs can he make?

　Ⓐ 1

　Ⓑ 2

　Ⓒ 8

　Ⓓ 10

13. Ella is making 3 batches of banana milkshakes. She needs $\frac{3}{4}$ gallon of milk for each batch. Her measuring cup holds $\frac{1}{4}$ gallon. How many times will she need to fill the measuring cup to make all 3 batches of milkshakes?

　Ⓐ 3

　Ⓑ 4

　Ⓒ 6

　Ⓓ 9

14. Darren cut a lemon pie into 8 equal slices. His friends ate some of the pie and now $\frac{5}{8}$ is left. Darren wants to put each slice of the leftover pie on its own plate. What part of the pie will he put on each plate?

　Ⓐ $\frac{1}{8}$

　Ⓑ $\frac{3}{8}$

　Ⓒ $\frac{5}{8}$

　Ⓓ $\frac{8}{8}$

15. Beth is putting liquid fertilizer on the plants in 4 flowerpots. Her measuring spoon holds $\frac{1}{8}$ teaspoon. The directions say to put $\frac{5}{8}$ teaspoon of fertilizer in each pot. How many times will Beth need to fill the measuring spoon to fertilize the plants in the 4 pots?

　Ⓐ 4

　Ⓑ 8

　Ⓒ 20

　Ⓓ 32

Name _____

Multiply a Fraction by a Whole Number Using Models

Essential Question How can you use a model to multiply a fraction by a whole number?

🔑 UNLOCK the Problem · REAL WORLD

Rafael practices the violin for $\frac{3}{4}$ hour each day. He has a recital in 3 days. How much time will he practice in 3 days?

> • How many equal groups of $\frac{3}{4}$ should you model?
>
> _____

🔒 Example 1 Use a model to multiply $3 \times \frac{3}{4}$.

Think: $3 \times \frac{3}{4}$ is 3 groups of $\frac{3}{4}$ of a whole. Shade the model to show 3 groups of $\frac{3}{4}$.

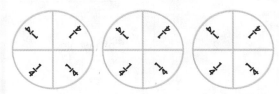

1 group of $\frac{3}{4}$ = _____

2 groups of $\frac{3}{4}$ = _____

3 groups of $\frac{3}{4}$ = _____

$3 \times \frac{3}{4}$ = _____

So, Rafael will practice for _____ hours in all.

> **MATHEMATICAL PRACTICES**
>
> **Math Talk** If you multiply $4 \times \frac{2}{6}$, is the product greater than or less than 4? **Explain**.

1. **Explain** how you can use repeated addition with the model to find the product of $3 \times \frac{3}{4}$.

_____

_____

2. Rafael's daily practice of $\frac{3}{4}$ hour is in sessions that last for $\frac{1}{4}$ hour each. **Describe** how the model shows the number of practice sessions Rafael has in 3 days.

_____

_____

🔒 Example 2 Use a pattern to multiply.

You know how to use a model and repeated addition to multiply a fraction by a whole number. Look for a pattern in the table to discover another way to multiply a fraction by a whole number.

Multiplication Problem		Whole Number (Number of Groups)	Fraction (Size of Groups)	Product
$\frac{1}{6}$ $\frac{1}{6}$ $\frac{1}{6}$ $\frac{1}{6}$ $\frac{1}{6}$ $\frac{1}{6}$ / $\frac{1}{6}$ $\frac{1}{6}$ $\frac{1}{6}$ $\frac{1}{6}$ $\frac{1}{6}$ $\frac{1}{6}$	$2 \times \frac{1}{6}$	2	$\frac{1}{6}$ of a whole	$\frac{2}{6}$
$\frac{1}{6}$ $\frac{1}{6}$ $\frac{1}{6}$ $\frac{1}{6}$ $\frac{1}{6}$ $\frac{1}{6}$ / $\frac{1}{6}$ $\frac{1}{6}$ $\frac{1}{6}$ $\frac{1}{6}$ $\frac{1}{6}$ $\frac{1}{6}$	$2 \times \frac{2}{6}$	2	$\frac{2}{6}$ of a whole	$\frac{4}{6}$
$\frac{1}{6}$ $\frac{1}{6}$ $\frac{1}{6}$ $\frac{1}{6}$ $\frac{1}{6}$ $\frac{1}{6}$ / $\frac{1}{6}$ $\frac{1}{6}$ $\frac{1}{6}$ $\frac{1}{6}$ $\frac{1}{6}$ $\frac{1}{6}$	$2 \times \frac{3}{6}$	2	$\frac{3}{6}$ of a whole	$\frac{6}{6}$

When you multiply a fraction by a whole number, the numerator

in the product is the product of the _____ and the

_____ of the fraction. The denominator in the product

is the same as the _____ of the fraction.

---

3. **Summarize** How do you multiply a fraction by a whole number without using a model or repeated addition?

_____

_____

4. **Describe** two different ways to find the product of $4 \times \frac{2}{3}$.

_____

_____

_____

Name _____

Multiply a Fraction or Mixed Number by a Whole Number

Essential Question How can you multiply a fraction by a whole number to solve a problem?

🔑 UNLOCK the Problem REAL WORLD

Christina is planning a dance routine. At the end of each measure of music, she will make a $1\frac{1}{4}$ turn. How many turns will she make after the first 3 measures of music?

You can multiply a mixed number by a whole number.

🔒 Example

STEP 1 Write and solve an equation.

$3 \times 1\frac{1}{4} = 3 \times$ ____ = ____   Write $1\frac{1}{4}$ as a fraction. Multiply.

STEP 2 Write the product as a mixed number.

$\frac{15}{4} = \frac{1}{4} + \frac{1}{4} + \frac{1}{4} + \frac{1}{4} +$ ____ + ____ + ____ + ____ + ____ + ____ + ____ + ____ + ____ + ____ + ____

$\underbrace{\qquad\qquad\qquad}_{1} + \underbrace{\qquad\qquad\qquad}_{1} + \cdot \underbrace{\qquad\qquad\qquad}_{1} + \frac{1}{4} + \frac{1}{4} + \frac{1}{4}$

= ____ + ____   Combine the wholes. Then combine the remaining parts.

= ____   Write the mixed number.

So, Christina will make _____ turns.

- Will Christina make more or less than $1\frac{1}{4}$ turns in 3 measures of music?

  _____

- What operation will you use to solve the problem?

  _____

Math Talk
MATHEMATICAL PRACTICES
Explain how writing the mixed number as a fraction in Step 2 is related to division.

1. If you multiply $3 \times \frac{1}{4}$, is the product greater than or less than 3? **Explain.**

   _____

   _____

2. **Explain** how you can tell that $3 \times 1\frac{1}{4}$ is greater than 3 without finding the exact product.

   _____

   _____

Chapter 8 329

Rename Mixed Numbers and Fractions You can use multiplication and division to rename fractions and mixed numbers.

 Write $8\frac{1}{5}$ as a fraction.

$8\frac{1}{5} = 8 + \frac{1}{5}$

$= (8 \times \underline{\hspace{1cm}}) + \frac{1}{5}$ Use the Identity Property of Multiplication.

$= \left(8 \times \underline{\hspace{0.8cm}}\right) + \frac{1}{5}$ Rename 1.

$= \underline{\hspace{0.8cm}} + \underline{\hspace{0.8cm}}$ Multiply.

$= \underline{\hspace{0.8cm}}$ Add.

 Write $\frac{32}{5}$ as a mixed number.

Find how many groups of $\frac{5}{5}$ are in $\frac{32}{5}$.

• Divide 32 by 5.

• The quotient is the number of wholes in $\frac{32}{5}$.

• The remainder is the number of fifths left over.

$5\overline{)32}$ r

There are 6 groups of $\frac{5}{5}$, or 6 wholes. There are 2 fifths, or $\frac{2}{5}$ left over.

$\frac{32}{5} = \underline{\hspace{0.8cm}}\underline{\hspace{0.8cm}}$

Try This! Find $5 \times 2\frac{2}{3}$. Write the product as a mixed number.

$5 \times 2\frac{2}{3} = 5 \times \underline{\hspace{1.5cm}}$ Write $2\frac{2}{3}$ as a fraction.

$= \underline{\hspace{1.5cm}}$ Multiply.

$= \underline{\hspace{1.5cm}}$ Divide the numerator by 3.

3. **Explain** why your solution to $5 \times 2\frac{2}{3} = 13\frac{1}{3}$ is reasonable.

_____

_____

4. **Sense or Nonsense?** To find $5 \times 2\frac{2}{3}$, Dylan says he can find $(5 \times 2) + \left(5 \times \frac{2}{3}\right)$. Does this make sense? **Explain.**

_____

_____

Name _____

Problem Solving • Comparison Problems with Fractions

Essential Question How can you use the strategy *draw a diagram* to solve comparison problems with fractions?

UNLOCK the Problem REAL WORLD

The deepest part of the Grand Canyon is about $1\frac{1}{6}$ miles deep. The deepest part of the ocean is located in the Mariana Trench, in the Pacific Ocean. The deepest part of the ocean is almost 6 times as deep as the deepest part of the Grand Canyon. About how deep is the deepest part of the ocean?

Read the Problem	Solve the Problem
What do I need to find? I need to find _____ _____ _____	Draw a bar model. Compare the depth of the deepest part of the Grand Canyon and the deepest part of the ocean, in miles. _____ $\boxed{1\frac{1}{6}}$ _____ 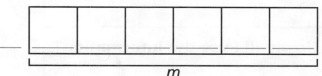 m
What information do I need to use? The deepest part of the Grand Canyon is about _____ miles deep. The deepest part of the ocean is about _____ times as deep.	Write an equation and solve. m is the deepest part of _____, in miles. $m =$ _____ ▢ _____ Write an equation. $m =$ _____ ▢ _____ Write the mixed number as a fraction. $m =$ _____ Multiply.
How will I use the information? I can _____ to compare the depths.	$m =$ _____ Write the fraction as a whole number.

So, the deepest part of the ocean is about _____ miles deep.

🔑 Try Another Problem

Mountains are often measured by the distance they rise above sea level. Mount Washington rises more than $1\frac{1}{10}$ miles above sea level. Mount Everest rises about 5 times as high. About how many miles above sea level does Mount Everest rise?

Read the Problem	Solve the Problem
What do I need to find?	
What information do I need to use?	
How will I use the information?	

So, Mount Everest rises about _____ miles above sea level.

- How did drawing a diagram help you solve the problem?

_____

_____

Math Talk MATHEMATICAL PRACTICES
Explain how you could use the strategy *act it out* to find the height of Mount Everest.

Name _____

Share and Show MATH BOARD

1. Komodo dragons are the heaviest lizards on earth.
   A baby Komodo dragon is $1\frac{1}{4}$ feet long when it
   hatches. Its mother is 6 times as long. How long
   is the mother?

   First, draw a bar model to show the problem.

   SHOW YOUR WORK

   Then, write the equation you need to solve.

   _____

   Finally, find the length of the mother Komodo dragon.

   The mother Komodo dragon is _____ feet long.

2. **H.O.T.** What if a male Komodo dragon is 7 times as long
   as the baby Komodo dragon? How long is the male? How
   much longer is the male than the mother?

   _____

✓ 3. The smallest hummingbird is the Bee hummingbird. It has a
   mass of about $1\frac{1}{2}$ grams. A Rufous hummingbird's mass is
   3 times the mass of the Bee hummingbird. What is the
   mass of a Rufous hummingbird?

   _____

✓ 4. Sloane needs $\frac{3}{4}$ hour to drive to her grandmother's house.
   It takes her 5 times as long to drive to her cousin's house.
   How long does it take to drive to her cousin's house?

   _____

On Your Own . . . .

Choose a STRATEGY

Act It Out
Draw a Diagram
Find a Pattern
Make a Table or List
Solve a Simpler Problem

Use the table for 5 and 6.

Payton has a variety of flowers in her garden. The table shows the average height of the flowers.

Flower	Height
tulip	$1\frac{1}{4}$ feet
daisy	$2\frac{1}{2}$ feet
tiger lily	$3\frac{1}{3}$ feet
sunflower	$7\frac{3}{4}$ feet

SHOW YOUR WORK

5. What is the difference between the tallest flower and the shortest flower in Payton's garden?

6. ⬤Write Math⬤ Payton says her average sunflower is 7 times the height of her average tulip. Do you agree or disagree with her statement? **Explain** your reasoning.

7. ⬤H.O.T.⬤ Miguel ran $1\frac{3}{10}$ miles on Monday. He wants to increase the distance he runs each day, so that on Friday he runs 3 times the distance he did on Monday. How far will Miguel run on Friday?

8. **Test Prep** Jack bought $1\frac{3}{4}$ pounds of cheese for a platter. He bought 3 times as much deli meat as cheese. How many pounds of deli meat did Jack buy?

   Ⓐ $1\frac{5}{4}$ pounds

   Ⓑ 5 pounds

   Ⓒ $5\frac{1}{4}$ pounds

   Ⓓ 7 pounds

FOR MORE PRACTICE:
Standards Practice Book, pp. P165–P166

Name _____

▶ **Vocabulary**

Choose the best term from the box.

Vocabulary
fraction
multiple
product

1. A _____ can name part of a whole or part of a group. (p. 316)

2. A _____ of a number is the product of the number and a counting number. (p. 316)

▶ **Concepts and Skills**

List the next four multiples of the unit fraction.

3. $\frac{1}{8}$, ▢ , ▢ , ▢ , ▢

4. $\frac{1}{4}$, ▢ , ▢ , ▢ , ▢

Write the fraction as a product of a whole number and a unit fraction.

5. $\frac{7}{12} =$ _____

6. $\frac{4}{12} =$ _____

7. $\frac{5}{4} =$ _____

List the next four multiples of the fraction.

8. $\frac{3}{10}$, ▢ , ▢ , ▢ , ▢

9. $\frac{2}{3}$, ▢ , ▢ , ▢ ,

Write the product as the product of a whole number and a unit fraction.

10. $3 \times \frac{2}{4} =$ _____

11. $2 \times \frac{3}{5} =$ _____

12. $4 \times \frac{2}{3} =$ _____

Multiply.

13. $5 \times \frac{7}{10} =$ _____

14. $4 \times \frac{3}{4} =$ _____

15. $3 \times \frac{8}{12} =$ _____

Multiply. Write the product as a mixed number.

16. $3 \times 1\frac{1}{8} =$ _____

17. $2 \times 2\frac{1}{5} =$ _____

18. $8 \times 1\frac{3}{5} =$ _____

GO Online — Assessment Options **Chapter Test**

Fill in the bubble completely to show your answer.

19. Bryson has soccer practice for $2\frac{1}{4}$ hours 2 times a week. How much time does Bryson spend at soccer practice in 1 week?

Ⓐ 2 hours

Ⓑ 4 hours

Ⓒ $4\frac{2}{4}$ hours

Ⓓ $8\frac{2}{4}$ hours

20. Nigel cut a loaf of bread into 12 equal slices. His family ate some of the bread and now $\frac{5}{12}$ is left. Nigel wants to put each of the leftover slices in its own bag. How many bags does Nigel need?

Ⓐ 5

Ⓑ 7

Ⓒ 12

Ⓓ 17

21. Micala made a list of some multiples of $\frac{3}{5}$. Which could be Micala's list?

Ⓐ $\frac{3}{5}, \frac{9}{5}, \frac{12}{5}, \frac{19}{5}$

Ⓑ $\frac{3}{5}, \frac{6}{10}, \frac{9}{15}, \frac{12}{20}$

Ⓒ $\frac{1}{5}, \frac{3}{5}, \frac{6}{5}, \frac{9}{5}$

Ⓓ $\frac{3}{5}, \frac{6}{5}, \frac{9}{5}, \frac{12}{5}$

22. Lincoln spent $1\frac{1}{4}$ hours reading a book. Phoebe spent 3 times as much time as Lincoln reading a book. How much time did Phoebe spend reading?

Ⓐ $1\frac{1}{16}$ hours

Ⓑ $3\frac{1}{4}$ hours

Ⓒ $3\frac{3}{4}$ hours

Ⓓ $4\frac{1}{4}$ hours

Name _____

Fill in the bubble completely to show your answer.

23. Griffin used a number line to write the multiples of $\frac{3}{8}$. Which multiple on the number line shows the product $2 \times \frac{3}{8}$?

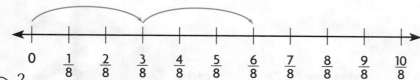

$0 \quad \frac{1}{8} \quad \frac{2}{8} \quad \frac{3}{8} \quad \frac{4}{8} \quad \frac{5}{8} \quad \frac{6}{8} \quad \frac{7}{8} \quad \frac{8}{8} \quad \frac{9}{8} \quad \frac{10}{8}$

(A) $\frac{2}{8}$

(B) $\frac{3}{8}$

(C) $\frac{6}{8}$

(D) $\frac{9}{8}$

24. Serena's rabbit weighs $3\frac{1}{2}$ pounds. Jarod's rabbit weighs 3 times as much as Serena's rabbit. How much does Jarod's rabbit weigh?

(A) $3\frac{1}{6}$ pounds

(B) $7\frac{1}{6}$ pounds

(C) $9\frac{1}{2}$ pounds

(D) $10\frac{1}{2}$ pounds

25. Jacadi is setting up a tent. Each section of a tent pole is $\frac{2}{3}$ yard long. She needs 4 sections to make 1 pole. How long is 1 tent pole?

(A) $\frac{12}{3}$ yards

(B) $\frac{8}{3}$ yards

(C) 8 yards

(D) $\frac{4}{3}$ yards

▶ Constructed Response

26. Oliver has music lessons Monday, Wednesday, and Friday. Each lesson is $\frac{3}{4}$ hour. Oliver says he will have lessons for $2\frac{1}{2}$ hours this week. Do you agree or disagree? **Explain** your reasoning.

_____

_____

▶ Performance Task

27. The common snapping turtle is a freshwater turtle. It can grow to about $1\frac{1}{6}$ feet long. The leatherback sea turtle is the largest of all sea turtles. The average length of a leatherback is about 5 times as long as a common snapping turtle.

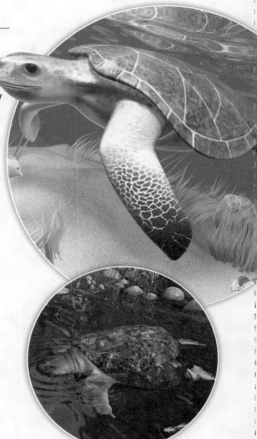

A Draw a diagram to compare the lengths of the turtles. Then write an equation to find the length of a leatherback. **Explain** how the diagram helps you write the equation.

_____

_____

B About how long is the leatherback sea turtle? _____

C A loggerhead sea turtle is about 3 times as long as the common snapping turtle. How long is the loggerhead? **Explain** your answer.

_____

_____

_____

_____

Show What You Know ✓

Check your understanding of important skills.

Name _____

▶ **Count Coins** Find the total value.

1. Total value: _____

2. Total value: _____

▶ **Equivalent Fractions**

Write two equivalent fractions for the picture.

3.

_____

4.

_____

▶ **Fractions with Denominators of 10**

Write a fraction for the words. You may draw a picture.

5. three tenths _____

6. six tenths _____

7. eight tenths _____

8. nine tenths _____

MATH DETECTIVE

WITH

CARMEN SANDIEGO

The Hudson River Science Barge, docked near New York City, provides a demonstration of how renewable energy can be used to produce food for large cities. Vegetables grown on the barge require _____ of the water needed by field crops. Be a Math Detective. Use these clues to find the fraction and decimal for the missing amount.

• The number is less than one and has two decimal places.
• The digit in the hundredths place has a value of $\frac{5}{100}$.
• The digit in the tenths place has a value of $\frac{2}{10}$.

Vocabulary Builder

▶ **Visualize It**

Complete the Semantic Map by using words with a ✓.

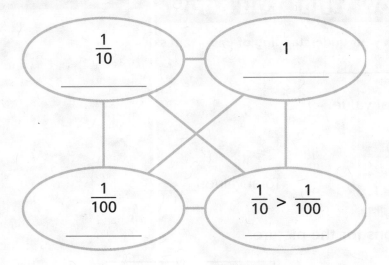

Review Words

✓ compare
  equivalent fractions
  fraction
  place value
✓ whole

Preview Words

  decimal
  decimal point
  equivalent decimals
✓ hundredth
✓ tenth

▶ **Understand Vocabulary**

Draw a line to match each word with its definition.

Word	Definition

Word

1. decimal

2. decimal point

3. tenth

4. hundredth

5. equivalent decimals

Definition

- Two or more decimals that name the same amount

- One part out of one hundred equal parts

- A number with one or more digits to the right of the decimal point

- One part out of ten equal parts

- A symbol used to separate dollars from cents in money amounts and to separate the ones and the tenths places in decimals

GO Online • eStudent Edition • Multimedia eGlossary

Name _____

Relate Tenths and Decimals

Essential Question How can you record tenths as fractions and decimals?

🔑 UNLOCK the Problem REAL WORLD

Ty is reading a book about metamorphic rocks. He has read $\frac{7}{10}$ of the book. What decimal describes the part of the book Ty has read?

A **decimal** is a number with one or more digits to the right of the **decimal point**. You can write tenths and hundredths as fractions or decimals.

🔒 One Way Use a model and a place-value chart.

Fraction

Shade $\frac{7}{10}$ of the model.

Think: The model is divided into 10 equal parts. Each part represents one **tenth**.

Write: _____

Read: seven tenths

Decimal

$\frac{7}{10}$ is 7 tenths.

Ones	.	Tenths	Hundredths
	.		

↑ decimal point

Write: _____

Read: _____

🔒 Another Way Use a number line.

Label the number line with decimals that are equivalent to the fractions. Locate the point $\frac{7}{10}$.

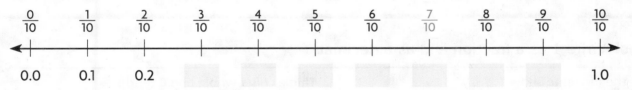

$\frac{0}{10}$ $\frac{1}{10}$ $\frac{2}{10}$ $\frac{3}{10}$ $\frac{4}{10}$ $\frac{5}{10}$ $\frac{6}{10}$ $\frac{7}{10}$ $\frac{8}{10}$ $\frac{9}{10}$ $\frac{10}{10}$

0.0   0.1   0.2   1.0

_____ names the same amount as $\frac{7}{10}$.

So, Ty read 0.7 of the book.

Math Talk MATHEMATICAL PRACTICES **Explain** how the size of one whole is related to the size of one tenth.

• How can you write 0.1 as a fraction? **Explain.**

_____

Tara rode her bicycle $1\frac{6}{10}$ miles. What decimal describes how far she rode her bicycle?

You have already written a fraction as a decimal. You can also write a mixed number as a decimal.

🔑 One Way Use a model and a place-value chart.

Fraction

Shade $1\frac{6}{10}$ of the model.

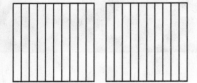

Write: _____

Read: one and six tenths

Decimal

$1\frac{6}{10}$ is 1 whole and 6 tenths.

Think: Use the ones place to record wholes.

Ones	.	Tenths	Hundredths
	.		

Write: _____

Read: _____

🔑 Another Way Use a number line.

Label the number line with equivalent mixed numbers and decimals. Locate the point $1\frac{6}{10}$.

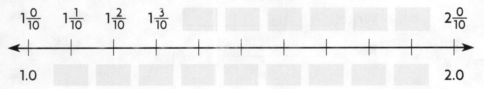

$1\frac{0}{10}$ $1\frac{1}{10}$ $1\frac{2}{10}$ $1\frac{3}{10}$ $2\frac{0}{10}$

1.0   2.0

_____ names the same amount as $1\frac{6}{10}$.

So, Tara rode her bicycle _____ miles.

Try This! Write 1 as a fraction and as a decimal.

Shade the model to show 1.

Fraction: _____

Think: 1 is 1 whole and 0 tenths.

Ones	.	Tenths	Hundredths
	.		

Decimal: _____

Name _____

Share and Show

1. Write five tenths as a fraction and as a decimal.

   Fraction: _____   Decimal: _____

Ones	.	Tenths	Hundredths
	.		

Write the fraction or mixed number and the decimal shown by the model.

2.

   _____

3.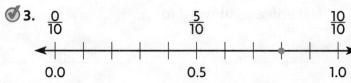

   _____

Math Talk MATHEMATICAL PRACTICES

How can you write $1\frac{3}{10}$ as a decimal? **Explain.**

On Your Own

Write the fraction or mixed number and the decimal shown by the model.

4.

   _____

5. $1\frac{0}{10}$ _____ $1\frac{5}{10}$

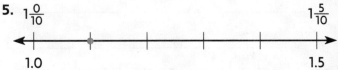

6.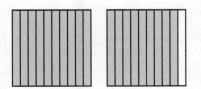

   _____

7. $3\frac{0}{10}$ _____ $3\frac{5}{10}$ _____ $4\frac{0}{10}$

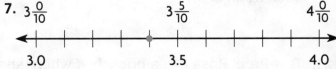

Practice: Copy and Solve Write the fraction or mixed number as a decimal.

8. $5\frac{9}{10}$

9. $\frac{1}{10}$

10. $\frac{7}{10}$

11. $8\frac{9}{10}$

12. $\frac{6}{10}$

13. $6\frac{3}{10}$

14. $\frac{5}{10}$

15. $9\frac{7}{10}$

© Houghton Mifflin Harcourt Publishing Company

Chapter 9 • Lesson 1 345

Problem Solving REAL WORLD

Use the table for 16–19.

16. What part of the rocks listed in the table are igneous? Write your answer as a decimal.

_____

17. Sedimentary rocks make up what part of Ramon's collection? Write your answer as a fraction and in word form.

_____

18. What part of the rocks listed in the table are metamorphic? Write your answer as a fraction and as a decimal.

_____

19. ☀ H.O.T. **What's the Error?** Niki wrote the following sentence in her report: "Metamorphic rocks make up 2.0 of Ramon's rock collection." Describe her error.

_____

_____

20. ☀ H.O.T. Josh paid for three books with two $20 bills. He received $1 in change. Each book was the same price. How much did each book cost?

_____

_____

21. **Test Prep** Rosa has a bookshelf where she stores her book collection. Four tenths of her books are mystery books. What is this amount written as a decimal?

  Ⓐ 40.0   Ⓒ 0.4

  Ⓑ 4.0   Ⓓ 0.04

Ramon's Rock Collection

Name	Type
Basalt	Igneous
Rhyolite	Igneous
Granite	Igneous
Peridotite	Igneous
Scoria	Igneous
Shale	Sedimentary
Limestone	Sedimentary
Sandstone	Sedimentary
Mica	Metamorphic
Slate	Metamorphic

▲ Granite– Igneous

▲ Mica–Metamorphic

▲ Sandstone– Sedimentary

SHOW YOUR WORK

© Houghton Mifflin Harcourt Publishing Company

FOR MORE PRACTICE:
Standards Practice Book, pp. P171–P172

Name _____

Relate Hundredths and Decimals

Essential Question How can you record hundredths as fractions and decimals?

🔑 UNLOCK the Problem REAL WORLD

In the 2008 Summer Olympic Games, the winning time in the men's 100-meter butterfly race was only $\frac{1}{100}$ second faster than the second-place time. What decimal represents this fraction of a second?

You can write hundredths as fractions or decimals.

• Circle the numbers you need to use.

🔓 One Way Use a model and a place-value chart.

Fraction	Decimal

Fraction

Shade $\frac{1}{100}$ of the model.

Think: The model is divided into 100 equal parts. Each part represents one **hundredth**.

Write: _____

Read: one hundredth

Decimal

Complete the place-value chart. $\frac{1}{100}$ is 1 hundredth.

Ones	.	Tenths	Hundredths
0	.	0	1

Write: _____

Read: one hundredth

🔓 Another Way Use a number line.

Label the number line with equivalent decimals. Locate the point $\frac{1}{100}$.

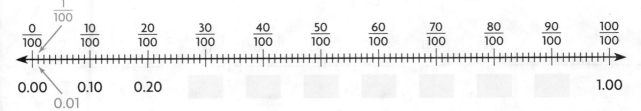

$\frac{1}{100}$

| $\frac{0}{100}$ | $\frac{10}{100}$ | $\frac{20}{100}$ | $\frac{30}{100}$ | $\frac{40}{100}$ | $\frac{50}{100}$ | $\frac{60}{100}$ | $\frac{70}{100}$ | $\frac{80}{100}$ | $\frac{90}{100}$ | $\frac{100}{100}$ |

0.00   0.10   0.20   1.00
0.01

_____ names the same amount as $\frac{1}{100}$.

So, the winning time was _____ second faster.

Math Talk MATHEMATICAL PRACTICES
Explain how the size of one tenth is related to the size of one hundredth.

Alicia won her 400-meter freestyle race by $4\frac{25}{100}$ seconds. How can you write this mixed number as a decimal?

🔑 One Way Use a model and a place-value chart.

Mixed Number

Shade the model to show $4\frac{25}{100}$.

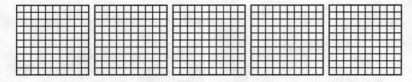

Write: _____

Read: four and twenty-five hundredths

_____

Decimal

Complete the place-value chart.

Think: Look at the model above. $4\frac{25}{100}$ is 4 wholes and 2 tenths 5 hundredths.

Ones	.	Tenths	Hundredths
	.		

Write: _____

Read: _____

🔑 Another Way Use a number line.

Label the number line with equivalent mixed numbers and decimals. Locate the point $4\frac{25}{100}$.

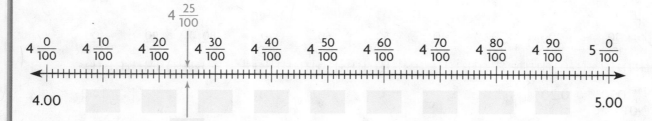

_____ names the same amount as $4\frac{25}{100}$.

So, Alicia won her race by _____ seconds.

Name _____

Share and Show

1. Shade the model to show $\frac{31}{100}$.

   Write the amount as a decimal. _____

Ones	.	Tenths	Hundredths
	.		

Write the fraction or mixed number and the decimal shown by the model.

 2. _____

3. _____ _____

 4.

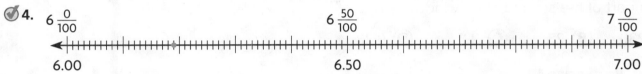
6 $\frac{0}{100}$ 6 $\frac{50}{100}$ 7 $\frac{0}{100}$

6.00   6.50   7.00

_____

Math Talk MATHEMATICAL PRACTICES
Are 0.5 and 0.50 equivalent? Explain.

On Your Own

Write the fraction or mixed number and the decimal shown by the model.

5. _____

6. _____

7.

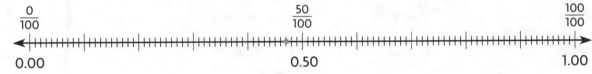
$\frac{0}{100}$ $\frac{50}{100}$ $\frac{100}{100}$

0.00   0.50   1.00

_____   _____

Practice: Copy and Solve Write the fraction or mixed number as a decimal.

8. $\frac{9}{100}$ 9. $4\frac{55}{100}$ 10. $\frac{10}{100}$ 11. $9\frac{33}{100}$ 12. $\frac{92}{100}$ 13. $14\frac{16}{100}$

Problem Solving REAL WORLD

14. **H.O.T.** Shade the grids to show three different ways to represent $\frac{16}{100}$ using models.

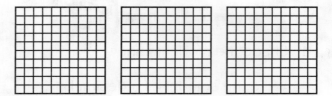

15. **Write Math** ▶ **Explain** how one whole, one tenth, and one hundredth are related.

_____

_____

16. **Test Prep** A stained glass window has 100 same-size squares. If 23 squares are red, what part of the stained glass window is red?

(A) 230   (B) 23   (C) 2.3   (D) 0.23

Sense or Nonsense?

17. The Memorial Library is 0.3 mile from school. Whose statement makes sense? Whose statement is nonsense? **Explain** your reasoning.

Gabe said he was going to walk 3 tenths mile to the Memorial Library after school.

Tara said she was going to walk 3 miles to the Memorial Library after school.

_____

_____

_____

_____

FOR MORE PRACTICE:
Standards Practice Book, pp. P173–P174

Name _____

Equivalent Fractions and Decimals

Essential Question How can you record tenths and hundredths as fractions and decimals?

🔑 UNLOCK the Problem REAL WORLD

Daniel spent a day hiking through a wildlife preserve. During the first hour of the hike, he drank $\frac{6}{10}$ liter of water. How many hundredths of a liter did he drink?

- Underline what you need to find.
- How can you represent hundredths?

_____

🔑 One Way Write $\frac{6}{10}$ as an equivalent fraction with a denominator of 100.

MODEL

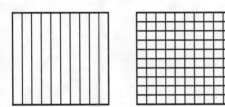

$\frac{6}{10} = \boxed{}{100}$

RECORD

$$\frac{6}{10} = \frac{6 \times \boxed{}}{10 \times \boxed{}} = \frac{\boxed{}}{100}$$

🔑 Another Way Write $\frac{6}{10}$ as a decimal.

Think: 6 tenths is the same as 6 tenths 0 hundredths.

Ones	.	Tenths	Hundredths

So, Daniel drank _____, or _____ liter of water.

Math Talk MATHEMATICAL PRACTICES
Explain how you can write 0.2 as hundredths.

- **Explain** why 6 tenths is equivalent to 60 hundredths.

_____

_____

Jasmine collected 0.30 liter of water in a jar during a rainstorm. How many tenths of a liter did she collect?

Equivalent decimals are decimals that name the same amount. You can write 0.30 as a decimal that names tenths.

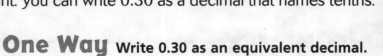

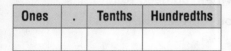 **One Way** Write 0.30 as an equivalent decimal.

Show 0.30 in the place-value chart.

Ones	.	Tenths	Hundredths

Think: There are no hundredths.

0.30 is equivalent to _____ tenths.

Write 0.30 as _____.

 **Another Way** Write 0.30 as a fraction with a denominator of 10.

STEP 1 Write 0.30 as a fraction.

0.30 is _____ hundredths.

30 hundredths written as a fraction is _____.

STEP 2 Write $\frac{30}{100}$ as an equivalent fraction with a denominator of 10.

Think: 10 is a common factor of the numerator and the denominator.

$$\frac{30}{100} = \frac{30 \div \boxed{}}{100 \div \boxed{}} = \frac{\boxed{}}{10}$$

So, Jasmine collected _____, or _____ liter of water.

Share and Show . . . .

1. Write $\frac{4}{10}$ as hundredths.

Write $\frac{4}{10}$ as an equivalent fraction.

$$\frac{4}{10} = \frac{4 \times \boxed{}}{10 \times \boxed{}} = \frac{\boxed{}}{100}$$

Write $\frac{4}{10}$ as a decimal.

Ones	.	Tenths	Hundredths
	.		

Fraction: _____

Decimal: _____

Name _____

**Write the number as hundredths in fraction form
and decimal form.**

✓ 2. $\frac{7}{10}$

3. 0.5

4. $\frac{3}{10}$

**Write the number as tenths in fraction form
and decimal form.**

✓ 5. 0.40

6. $\frac{80}{100}$

7. $\frac{20}{100}$

On Your Own ....

Math Talk MATHEMATICAL PRACTICES Can you write 0.25 as tenths? Explain.

Practice: Copy and Solve Write the number as
hundredths in fraction form and decimal form.

8. $\frac{8}{10}$

9. $\frac{2}{10}$

10. 0.1

Practice: Copy and Solve Write the number as tenths in
fraction form and decimal form.

11. $\frac{60}{100}$

12. $\frac{90}{100}$

13. 0.70

H.O.T. Write the number as an equivalent mixed number
with hundredths.

14. $1\frac{4}{10}$

15. $3\frac{5}{10}$

16. $2\frac{9}{10}$

© Houghton Mifflin Harcourt Publishing Company

Problem Solving REAL WORLD

17. **H.O.T.** **What's the Error?** Carter says that 0.08 is equivalent to $\frac{8}{10}$. Describe and correct Carter's error.

_____

_____

_____

18. **Test Prep** David walked 0.5 kilometer from his home to the library. Which fraction is equivalent to 0.5?

Ⓐ $\frac{5}{100}$ Ⓒ $\frac{55}{100}$

Ⓑ $\frac{50}{100}$ Ⓓ $\frac{50}{10}$

Connect to Science

Inland Water

How many lakes and rivers does your state have? The U.S. Geological Survey defines inland water as water that is surrounded by land. The Atlantic Ocean, the Pacific Ocean, and the Great Lakes are not considered inland water.

19. **Write Math** Just over $\frac{2}{100}$ of the entire United States is inland water. Write $\frac{2}{100}$ as a decimal.

_____

20. Can you write 0.02 as tenths? **Explain.**

_____

_____

21. About 0.17 of the area of Rhode Island is inland water. Write 0.17 as a fraction.

_____

22. Louisiana's lakes and rivers cover about $\frac{1}{10}$ of the state. Write $\frac{1}{10}$ as hundredths in fraction form and decimal form.

_____

Relate Fractions, Decimals, and Money

Essential Question How can you relate fractions, decimals, and money?

🔓 UNLOCK the Problem · REAL WORLD

Together, Julie and Sarah have $1.00 in quarters. They want to share the quarters equally. How many quarters should each girl get? How much money is this?

 Use the model to relate money, fractions, and decimals.

4 quarters = 1 dollar = $1.00

$0.25 $0.25 $0.25 $0.25

1 quarter is $\frac{25}{100}$, or $\frac{1}{4}$ of a dollar.

2 quarters are $\frac{50}{100}$, $\frac{2}{4}$, or $\frac{1}{2}$ of a dollar.

$\frac{1}{2}$ of a dollar = $0.50, or 50 cents.

Circle the number of quarters each girl should get.

So, each girl should get 2 quarters, or $ _____.

🔓 Examples Use money to model decimals.

1 dollar	10 dimes = 1 dollar	100 pennies = 1 dollar

$1.00, or

_____ cents

1 dime = $\frac{1}{10}$, or 0.10 of a dollar

$ _____, or 10 cents

1 penny = $\frac{1}{100}$, or 0.01 of a dollar

$ _____, or 1 cent

Math Talk · MATHEMATICAL PRACTICES
If you have 68 pennies, what part of a dollar do you have? **Explain.**

Relate Money and Decimals Think of dollars as ones, dimes as tenths, and pennies as hundredths.

$1.56

Dollars	.	Dimes	Pennies
1	.	5	6

Think: $1.56 = 1 dollar and 56 pennies

There are 100 pennies in 1 dollar.
So, $1.56 = 156 pennies.

1.56 dollars

Ones	.	Tenths	Hundredths
1	.	5	6

Think: 1.56 = 1 one and 56 hundredths

There are 100 hundredths in 1 one.
So, 1.56 = 156 hundredths.

🔒 More Examples

Shade the decimal model to show the money amount. Then write the money amount and a fraction in terms of dollars.

A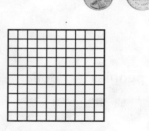

_____, or $\frac{21}{100}$ of a dollar

B

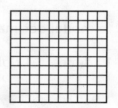 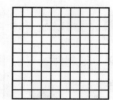

$1.46, or 1 $\frac{}{100}$ dollars

Try This! Complete the table to show how money, fractions, mixed numbers, and decimals are related.

$ Bills and Coins	Money Amount	Fraction or Mixed Number	Decimal
	$0.03		0.03
	$0.25	$\frac{25}{100}$, or $\frac{1}{4}$	
2 quarters 1 dime		$\frac{60}{100}$, or $\frac{6}{10}$	
2 $1 bills 5 nickels			

Math Talk

Would you rather have $0.25 or $\frac{3}{10}$ of a dollar? **Explain.**

356

Name _____

Share and Show

1. Write the amount of money as a decimal in terms of dollars.

   5 pennies = $\frac{5}{100}$ of a dollar = _____ of a dollar.

Write the total money amount. Then write the amount as a fraction or a mixed number and as a decimal in terms of dollars.

2.

   _____ _____ _____

3.

   _____ _____ _____

Write as a money amount and as a decimal in terms of dollars.

4. $\frac{92}{100}$ _____

5. $\frac{7}{100}$ _____

6. $\frac{16}{100}$ _____

7. $\frac{53}{100}$ _____

> **Math Talk** MATHEMATICAL PRACTICES
> **Explain** how $0.84 and $\frac{84}{100}$ of a dollar are related.

On Your Own

Write the total money amount. Then write the amount as a fraction or a mixed number and as a decimal in terms of dollars.

8.

   _____ _____ _____

9.

   _____ _____ _____

10.

   _____ _____ _____

11.

   _____ _____ _____

Write as a money amount and as a decimal in terms of dollars.

12. $\frac{27}{100}$ _____

13. $\frac{4}{100}$ _____

14. $\frac{75}{100}$ _____

15. $\frac{100}{100}$ _____

Write the money amount as a fraction in terms of dollars.

16. $0.68 _____   **17.** $0.20 _____   **18.** $0.89 _____   **19.** $0.47 _____

Write the total money amount. Then write the amount as a fraction and as a decimal in terms of dollars.

20. 1 quarter 6 dimes 8 pennies

_____ _____ _____

21. 3 dimes 5 nickels 20 pennies

_____ _____ _____

H.O.T. **Algebra** Complete to tell the value of each digit.

22. $1.05 = _____ dollar + _____ pennies,   1.05 = _____ one + _____ hundredths

23. $5.18 = _____ dollars + _____ dime + _____ pennies

5.18 = _____ ones + _____ tenth + _____ hundredths

Problem Solving > REAL WORLD

Use the table for 24–25.

24. The table shows the coins three students have. Write Nick's total amount as a fraction in terms of dollars.

_____

25. **H.O.T.** Kim spent $\frac{40}{100}$ of a dollar on a snack. Write as a money amount the amount she has left.

_____

Pocket Change				
Name	**Quarters**	**Dimes**	**Nickels**	**Pennies**
Kim	1	3	2	3
Tony	0	6	1	6
Nick	2	4	0	2

26. **Write Math** ▷ Travis has $\frac{1}{2}$ of a dollar. He has at least two different types of coins in his pocket. **Draw** two possible sets of coins that Travis could have.

27. Test Prep Mia has two dollars and fifteen cents. What decimal names this money amount in terms of dollars?

(A) 21.50   (C) 2.15

(B) 2.50   (D) 0.15

Problem Solving • Money

Essential Question How can you use the strategy *act it out* to solve problems that use money?

UNLOCK the Problem  REAL WORLD

Together, Marnie and Serena have $1.20. They want to share the money equally. How much money will each girl get?

Use the graphic organizer to solve the problem.

Read the Problem

What do I need to find?

I need to find the _____

_____

What information do I need to use?

I need to use the total amount, _____, and

divide the amount into _____ equal parts.

How will I use the information?

I will use coins to model the _____ and

act out the problem.

Solve the Problem

You can make $1.20 with 4 quarters

and 2 _____.

Circle the coins to show two sets with equal value.

So, each girl gets _____ quarters and

_____ dime. Each girl gets $_____.

• **Describe** another way you could act out the problem with coins.

_____

_____

🔑 Try Another Problem

Josh, Tom, and Chuck each have $0.40. How much money do they have together?

Read the Problem	Solve the Problem
What do I need to find?	
What information do I need to use?	
How will I use the information?	

- How can you solve the problem using dimes and nickels?

_____

_____

Math Talk MATHEMATICAL PRACTICES
What other strategy might you use to solve the problem? **Explain.**

Name _____

Share and Show

UNLOCK the Problem Tips
√ Circle the question.
√ Underline the important facts.
√ Cross out unneeded information.

1. Juan has $3.43. He is buying a paint brush that costs $1.21 to paint a model race car. How much will Juan have after he pays for the paint brush?

   First, use bills and coins to model $3.43.

   Next, you need to subtract. Remove bills and coins that have a value of $1.21. Mark Xs to show what you remove.

   Last, count the value of the bills and coins that are left. How much will Juan have left?

   _____

2. **H.O.T.** **What if** Juan has $3.43, and he wants to buy a paint brush that costs $2.28? How much money will Juan have left then? **Explain.**

   _____

   _____

   _____

   _____

   _____

3. Sophia has $2.25. She wants to give an equal amount to each of her 3 young cousins. How much will each cousin receive?

   _____

SHOW YOUR WORK

On Your Own

4. Marcus saves $13 each week. In how many weeks will he have saved at least $100?

_____

5. Hoshi has $50. Emily has $23 more than Hoshi. Karl has $16 less than Emily. How much money do they have all together?

_____

6. **Write Math** ▶ Four girls have $5.00 to share equally. How much money will each girl get? **Explain.**

_____

_____

_____

_____

7. **H.O.T.** **What if** four girls want to share $5.52 equally? How much money will each girl get? **Explain.**

_____

_____

_____

_____

_____

8. **Test Prep** Which does **not** represent the value of 3 quarters and 3 dimes?

(A) $1.05

(C) $1\frac{5}{10}$ dollars

(B) $1\frac{5}{100}$ dollars

(D) 1.05 dollars

MATHEMATICAL PRACTICES **Model • Reason • Make Sense**

Choose a STRATEGY

Act It Out

Draw a Diagram

Find a Pattern

Make a Table or List

Solve a Simpler Problem

SHOW YOUR WORK

© Houghton Mifflin Harcourt Publishing Company

FOR MORE PRACTICE:
Standards Practice Book, pp. P179–P180

Name _____

 **Mid-Chapter Checkpoint**

▶ **Vocabulary**

Choose the best term from the box to complete
the sentence.

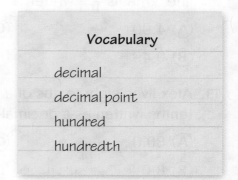

1. A symbol used to separate the ones and the tenths place is

   called a _____. (p. 343)

2. The number 0.4 is written as a _____. (p. 343)

3. A _____ is one of one hundred equal parts of a
   whole. (p. 347)

▶ **Concepts and Skills**

Write the fraction or mixed number and the decimal shown
by the model.

4.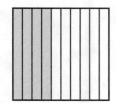

   _____   _____

5.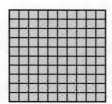

   _____   _____

Write the number as hundredths in fraction form
and decimal form.

6. $\frac{8}{10}$

   _____

7. 0.5

   _____

8. $\frac{6}{10}$

   _____

Write the fraction or mixed number as a money amount,
and as a decimal in terms of dollars.

9. $\frac{65}{100}$

   _____

10. $1\frac{48}{100}$

   _____

11. $\frac{4}{100}$

   _____

Fill in the bubble completely to show your answer.

12. Ken's turtle competed in a 0.50-meter race. His turtle had traveled $\frac{49}{100}$ meter when the winning turtle crossed the finish line. What is $\frac{49}{100}$ written as a decimal?

Ⓐ 4.90   Ⓒ 0.40

Ⓑ 0.49   Ⓓ 0.09

13. Alex lives eight tenths of a mile from Sarah. What is eight tenths written as a decimal?

Ⓐ 80.0   Ⓒ 0.8

Ⓑ 8.0   Ⓓ 0.08

14. Which two fractions are equivalent?

Ⓐ $\frac{1}{10}$ and $\frac{1}{100}$ Ⓒ $\frac{10}{100}$ and $\frac{1}{100}$

Ⓑ $\frac{10}{10}$ and $\frac{10}{100}$ Ⓓ $\frac{1}{10}$ and $\frac{10}{100}$

15. Elaine found the following in her pocket. How much money was in her pocket?

Ⓐ $1.04   Ⓒ $1.40

Ⓑ $1.30   Ⓓ $1.45

16. Three girls share $0.60. Each girl gets the same amount. How much money does each girl get?

Ⓐ $0.30   Ⓒ $0.20

Ⓑ $0.25   Ⓓ $0.02

17. The deli scale weighs meat and cheese in hundredths of a pound. Sam put $\frac{5}{10}$ pound of pepperoni on the deli scale. What weight does the deli scale show?

Ⓐ 0.05 pound   Ⓒ 0.50 pound

Ⓑ 0.10 pound   Ⓓ 5.0 pound

Add Fractional Parts of 10 and 100

Essential Question How can you add fractions when the denominators are 10 or 100?

🔓 UNLOCK the Problem REAL WORLD

The fourth grade classes are painting designs on tile squares to make a mural. Mrs. Kirk's class painted $\frac{3}{10}$ of the mural. Mr. Becker's class painted $\frac{21}{100}$ of the mural. What part of the mural is painted?

You know how to add fractions with parts that are the same size. You can use equivalent fractions to add fractions with parts that are not the same size.

🔑 Example 1 Find $\frac{3}{10} + \frac{21}{100}$.

STEP 1 Write $\frac{3}{10}$ and $\frac{21}{100}$ as a pair of fractions with a common denominator.

Think: 100 is a multiple of 10. Use 100 as the common denominator.

$$\frac{3}{10} = \frac{3 \times \boxed{}}{10 \times \boxed{}} = \frac{\boxed{}}{100}$$

Think: $\frac{21}{100}$ already has 100 in the denominator.

So, $\frac{\boxed{}}{100}$ of the mural is painted.

STEP 2 Add.

Think: Write $\frac{3}{10} + \frac{21}{100}$ using fractions with a common denominator.

$$\frac{30}{100} + \frac{21}{100} = \frac{\boxed{}}{100}$$

MATHEMATICAL PRACTICES

Math Talk When adding tenths and hundredths, can you always use 100 as a common denominator? **Explain.**

Try This! Find $\frac{4}{100} + \frac{1}{10}$.

A Write $\frac{1}{10}$ as $\frac{10}{100}$.

$$\frac{1}{10} = \frac{1 \times \boxed{}}{10 \times \boxed{}} = \frac{\boxed{}}{100}$$

B Add.

$$\frac{\boxed{}}{100} + \frac{10}{100} = \frac{\boxed{}}{100}$$

So, $\frac{4}{100} + \frac{10}{100} = \frac{14}{100}$.

🔑 Example   Add decimals.

Sean lives 0.5 mile from the store. The store is 0.25 mile from his grandmother's house. Sean is going to walk to the store and then to his grandmother's house. How far will he walk?

Find 0.5 + 0.25.

STEP 1 Write 0.5 + 0.25 as a sum of fractions.

Think: 0.5 is 5 tenths.   Think: 0.25 is 25 hundredths.

$$0.5 = \underline{}\qquad\qquad 0.25 = \underline{}$$

Write 0.5 + 0.25 as $\underline{}$ + $\underline{}$

STEP 2 Write $\frac{5}{10} + \frac{25}{100}$ as a sum of fractions with a common denominator.

Think: Use 100 as a common denominator. Rename $\frac{5}{10}$.

$$\frac{5}{10} = \frac{5 \times }{10 \times } = \frac{}{100}$$

Write $\frac{5}{10} + \frac{25}{100}$ as $\underline{}$ + $\underline{}$.

STEP 3 Add.

$$\frac{50}{100} + \frac{25}{100} = \underline{}$$

STEP 4 Write the sum as a decimal.

$$\frac{75}{100} = \underline{}$$

So, Sean will walk _____ mile.

MATHEMATICAL PRACTICES

Math Talk   Explain why you can think of $0.25 as either $\frac{1}{4}$ dollar or $\frac{25}{100}$ dollar.

Try This! Find $0.25 + $0.40.

$0.25 + $0.40 = _____

Remember

A money amount less than a dollar can be written as a fraction of a dollar.

Name _____

Share and Show   ·

1. Find $\dfrac{7}{10} + \dfrac{5}{100}$.

   Think: Write the addends as fractions with a common denominator.

   $$\dfrac{}{100} + \dfrac{}{100} = \dfrac{}{}$$

Find the sum.

2. $\dfrac{1}{10} + \dfrac{11}{100} =$ _____

3. $\dfrac{36}{100} + \dfrac{5}{10} =$ _____

4. $0.16 + $0.45 = $_____

5. $0.08 + $0.88 = $_____

On Your Own ·

6. $\dfrac{6}{10} + \dfrac{25}{100} =$ _____

7. $\dfrac{7}{10} + \dfrac{7}{100} =$ _____

8. $\dfrac{19}{100} + \dfrac{4}{10} =$ _____

9. $\dfrac{3}{100} + \dfrac{9}{10} =$ _____

10. $0.55 + $0.23 = $_____

11. $0.19 + $0.13 = $_____

 Algebra Write the number that makes the equation true.

12. $\dfrac{20}{100} + \dfrac{}{10} = \dfrac{60}{100}$

13. $\dfrac{2}{10} + \dfrac{}{100} = \dfrac{90}{100}$

Problem Solving REAL WORLD

Use the table for 14–17.

14. Dean selects Teakwood stones and Buckskin stones to pave a path in front of his house. How many meters long will each set of one Teakwood stone and one Buckskin stone be?

_____

15. The backyard patio at Nona's house is made from a repeating pattern of one Rose stone and one Rainbow stone. How many meters long is each pair of stones?

_____

16. **H.O.T.** For a stone path, Emily likes the look of a Rustic stone, then a Rainbow stone, and then another Rustic stone. How long will the three stones in a row be? **Explain.**

_____

_____

_____

17. **Write Math** Which two stones can you place end-to-end to get a length of 0.38 meters? **Explain** how you found your answer.

_____

_____

_____

Paving Stone Center	
Style	**Length (in meters)**
Rustic	$\frac{15}{100}$
Teakwood	$\frac{3}{10}$
Buckskin	$\frac{41}{100}$
Rainbow	$\frac{6}{10}$
Rose	$\frac{8}{100}$

SHOW YOUR WORK

18. **Test Prep** Which is the sum of $\frac{6}{10}$ and $\frac{2}{100}$?

Ⓐ $\frac{8}{10}$ Ⓒ $\frac{62}{10}$

Ⓑ $\frac{62}{100}$ Ⓓ $\frac{8}{100}$

FOR MORE PRACTICE:
Standards Practice Book, pp. P181–P182

Compare Decimals

Essential Question How can you compare decimals?

🔑 UNLOCK the Problem REAL WORLD

The city park covers 0.64 square mile. About 0.18 of the park is covered by water, and about 0.2 of the park is covered by paved walkways. Is more of the park covered by water or paved walkways?

- Cross out unnecessary information.
- Circle numbers you will use.
- What do you need to find?

_____

_____

🔒 One Way Use a model.

Shade 0.18.

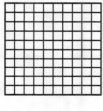

Shade 0.2.

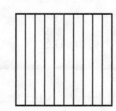

0.18 ◯ 0.2

🔒 Other Ways

A Use a number line.

Locate 0.18 and 0.2 on a number line.

Think: 2 tenths is equivalent to 20 hundredths.

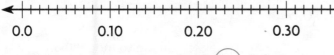

0.0   0.10   0.20   0.30   0.40   0.50

_____ is closer to 0, so 0.18 ◯ 0.2.

B Compare equal-size parts.

- 0.18 is _____ hundredths.

- 0.2 is 2 tenths, which is equivalent to _____ hundredths.

18 hundredths ◯ 20 hundredths, so 0.18 ◯ 0.2.

So, more of the park is covered by _____.

Math Talk MATHEMATICAL PRACTICES

How does the number of tenths in 0.18 compare to the number of tenths in 0.2? **Explain.**

Place Value You can compare numbers written as decimals by using place value. Comparing decimals is like comparing whole numbers. Always compare the digits in the greatest place-value position first.

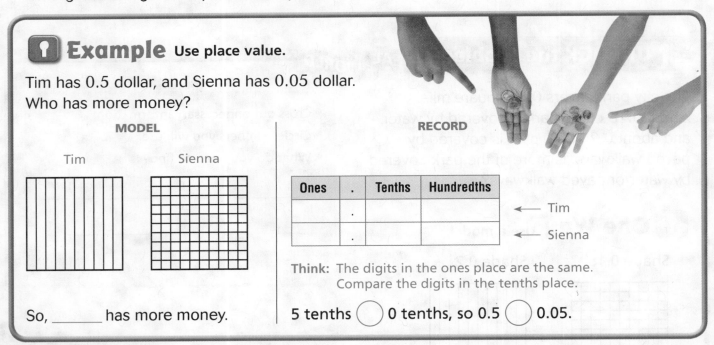

🔒 **Example** Use place value.

Tim has 0.5 dollar, and Sienna has 0.05 dollar. Who has more money?

MODEL

Tim   Sienna

RECORD

Ones	.	Tenths	Hundredths	
	.			← Tim
	.			← Sienna

Think: The digits in the ones place are the same. Compare the digits in the tenths place.

So, _____ has more money.

5 tenths ◯ 0 tenths, so 0.5 ◯ 0.05.

- Compare the size of 1 tenth to the size of 1 hundredth. How could this help you compare 0.5 and 0.05? **Explain.**

_____

_____

_____

Try This! Compare 1.3 and 0.6. Write <, >, or =.

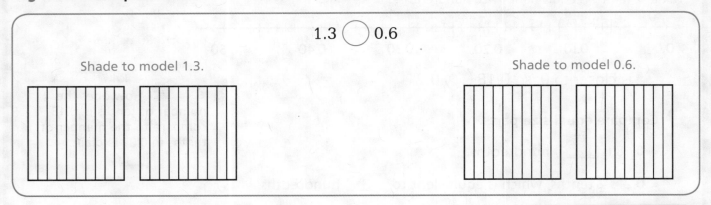

1.3 ◯ 0.6

Shade to model 1.3.

Shade to model 0.6.

MATHEMATICAL PRACTICES

Math Talk Explain how you could use place value to compare 1.3 and 0.6.

Name _____

Share and Show

1. Compare 0.39 and 0.42. Write <, >, or =.
   Shade the model to help.

   0.39 ◯ 0.42

   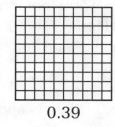   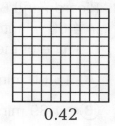

   0.39   0.42

Compare. Write <, >, or =.

2. 0.26 ◯ 0.23

Ones	.	Tenths	Hundredths
	.		
	.		

◉ 3. 0.7 ◯ 0.54

Ones	.	Tenths	Hundredths
	.		
	.		

4. 1.15 ◯ 1.3

Ones	.	Tenths	Hundredths
	.		
	.		

◉ 5. 4.5 ◯ 2.89

Ones	.	Tenths	Hundredths
	.		
	.		

MATHEMATICAL PRACTICES

Math Talk Can you compare 0.39 and 0.42 by comparing only the tenths? **Explain.**

On Your Own

Compare. Write <, >, or =.

6. 0.9 ◯ 0.81   7. 1.06 ◯ 0.6   8. 0.25 ◯ 0.3   9. 2.61 ◯ 3.29

10. 0.38 ◯ 0.83   11. 1.9 ◯ 0.99   12. 1.11 ◯ 1.41   13. 0.8 ◯ 0.80

 Compare. Write <, >, or =.

14. 0.30 ◯ $\frac{3}{10}$ 15. $\frac{4}{100}$ ◯ 0.2   16. 0.15 ◯ $\frac{1}{10}$ 17. $\frac{1}{8}$ ◯ 0.8

UNLOCK the Problem REAL WORLD

18. Ricardo and Brandon ran a 1500-meter race. Ricardo finished in 4.89 minutes. Brandon finished in 4.83 minutes. What was the time of the runner who finished first?

Ⓐ 15.00 minutes   Ⓒ 4.83 minutes

Ⓑ 4.89 minutes   Ⓓ Ricardo and Brandon tied for first.

a. What are you asked to find? _____

_____

b. What do you need to do to find the answer? _____

_____

c. Solve the problem.

d. Fill in the bubble for the correct answer choice above.

e. Look back. Does your answer make sense? Explain.

_____

_____

_____

_____

_____

_____

19. Which of the following is less than $3.68?

Ⓐ $3.97

Ⓑ $4.57

Ⓒ $3.59

Ⓓ $4.68

20. Which of the following is greater than 14.24?

Ⓐ 13.99

Ⓑ 14.19

Ⓒ 13.24

Ⓓ 14.34

FOR MORE PRACTICE:
Standards Practice Book, pp. P183–P184

Name _____

▶ **Vocabulary**

Choose the best term from the box.

Vocabulary
decimal
decimal point
hundredth
tenth

1. One of ten equal parts is one _____.
   (p. 343)

2. A _____ is a symbol used to separate dollars from cents in money amounts and to separate the ones and the tenths places in decimals. (p. 343)

3. A _____ is a number with one or more digits to the right of the decimal point. (p. 343)

▶ **Concepts and Skills**

Write the fraction and the decimal shown by the model.

4.

_____   _____

5.

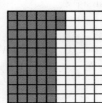

_____   _____

Write the number as hundredths in fraction form and decimal form.

6. $\frac{9}{10}$

_____

7. $\frac{3}{10}$

_____

8. 0.2

_____

Find the sum.

9. $\frac{5}{10} + \frac{30}{100} =$ _____

10. $\frac{6}{10} + \frac{4}{100} =$ _____

11. $0.24 + 0.1 =$ _____

Compare. Write <, >, or =.

12. $3.45 \bigcirc 3.54$

13. $1.7 \bigcirc 1.70$

14. $8.1 \bigcirc 8.01$

15. $\$4.25 \bigcirc \$3.75$

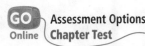
Assessment Options
Chapter Test

Fill in the bubble completely to show your answer.

16. Which fraction or mixed number and decimal is shown by the model?

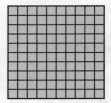 

Ⓐ $\frac{24}{100}$, 0.24

Ⓑ $1\frac{24}{100}$, 1.24

Ⓒ $1\frac{76}{100}$, 1.76

Ⓓ $1\frac{24}{10}$, 1.24

17. Bethany collected 0.7 inch of rain in her rain gauge. How many hundredths of an inch did she collect?

Ones	.	Tenths	Hundredths
	.		

Ⓐ $\frac{7}{100}$ Ⓒ $\frac{7}{10}$

Ⓑ $\frac{70}{100}$ Ⓓ $\frac{7}{1}$

18. Pam paid for her lunch with the amount of money shown below.

How much money did she spend?

Ⓐ $2\frac{62}{100}$ dollars   Ⓒ $2\frac{87}{100}$ dollars

Ⓑ $2\frac{77}{100}$ dollars   Ⓓ $3\frac{2}{100}$ dollars

Fill in the bubble completely to show your answer.

19. Carson shaded a model to represent the part of his book he read this weekend. Which decimal represents the part of the book he read?

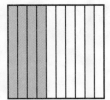

   Ⓐ 4.0

   Ⓑ 0.44

   Ⓒ 0.4

   Ⓓ 0.04

20. Christelle is making a doll house. The doll house is $\frac{6}{10}$ meter high without the roof. The roof is $\frac{15}{100}$ meter high. What will the height of the doll house be, with the roof?

   Ⓐ $\frac{21}{100}$ meter

   Ⓑ $\frac{75}{100}$ meter

   Ⓒ $1\frac{6}{100}$ meter

   Ⓓ $\frac{60}{100}$ meter

21. Amie has three quarters and one nickel. If she and three girls share the money equally, what will each person get?

   Ⓐ $0.10

   Ⓑ $0.15

   Ⓒ $0.20

   Ⓓ $0.25

▶ Constructed Response

22. There is $\frac{30}{100}$ of a liter of orange juice in one container and $\frac{5}{10}$ of a liter of pineapple juice in another container. If Mrs. Morales combines the two juices, how much orange-pineapple juice will she have? **Explain** how you found your answer.

_____

_____

_____

23. Write the amount of orange-pineapple juice as a decimal.

_____

▶ Performance Task

24. Luke lives 0.4 kilometer from a skating rink. Mark lives 0.25 kilometer from the skating rink.

A Who lives closer to the skating rink? **Explain**.

_____

_____

_____

B How can you write each distance as a fraction? **Explain**.

_____

_____

C Luke is walking to the skating rink to pick up a practice schedule. Then he is walking to Mark's house. Will he walk more than a kilometer or less than a kilometer? **Explain**.

_____

_____

_____

Geometry, Measurement, and Data

Understanding that geometric figures can be analyzed and classified based on their properties, such as having parallel sides, perpendicular sides, particular angle measures, and symmetry

Landscape architects can help design and plan outdoor spaces such as botanical gardens.

Project

Landscape Architects

When people who live and work in big cities take breaks, they leave their tall buildings to relax in patches of green. A city garden may be small, but it gives people a chance to enjoy the beauty of nature.

Get Started

Design a garden that covers a whole city block. Decide on features to have in your garden and where they will be located. Mark off parts of your garden for each feature. Then find the number of square units the feature covers and record it on the design. Use the Important Facts to help you.

Important Facts

Features of a City Garden

Benches			Snack bar
Flower garden			Spring bulb garden
Paths			Tree garden
Shrub garden			Waterfall and fountain

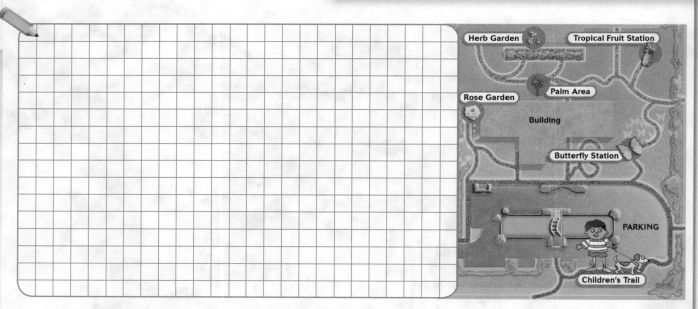

▲ This map is an example of how a city garden could be laid out.

Completed by _____

10 Two-Dimensional Figures

Show What You Know ✓

Check your understanding of important skills.

Name _____

▶ **Sides and Vertices** Write the number of vertices.

1.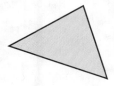

_____ vertices

2.

_____ vertices

3.

_____ vertices

▶ **Number of Sides** Write the number of sides.

4.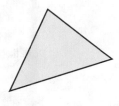

_____ sides

5.

_____ sides

6.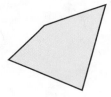

_____ sides

▶ **Geometric Patterns** Draw the next two shapes in the pattern.

7.

MATH DETECTIVE

WITH **CARMEN SANDIEGO**™

The Isle of Wight Natural History Centre, off the coast of England, has shells of every size, shape, and color. Many shells have symmetry. Be a Math Detective. Investigate this shell. Describe its shape in geometric terms. Then determine whether this shell has line symmetry.

GO Online   Assessment Options: **Soar to Success Math**

Vocabulary Builder

▶ **Visualize It** •

Complete the flow map by using the words with a ✓.

Geometry

What is it?

What are some examples?

Review Words

✓ polygon

✓ triangle

✓ quadrilateral

Preview Words

acute angle

acute triangle

line

line segment

line symmetry

obtuse angle

obtuse triangle

parallel lines

parallelogram

perpendicular lines

ray

right angle

right triangle

straight angle

▶ **Understand Vocabulary** • • • • • • • • • • • • • • • • •

Complete the sentences by using preview words.

1. A shape has _____ if it can be folded about a line so that its two parts match exactly.

2. A figure that has no endpoints is called a _____.

3. A figure that has two endpoints is called a _____.

4. _____ are lines that never cross.

5. When two lines cross to form a square corner, the lines are _____.

GO Online • eStudent Edition • Multimedia eGlossary

Lines, Rays, and Angles

Essential Question How can you identify and draw points, lines, line segments, rays, and angles?

🔑 UNLOCK the Problem REAL WORLD

Everyday things can model geometric figures. For example, the period at the end of this sentence models a point. A solid painted stripe in the middle of a straight road models a line.

Term and Definition	Draw It	Read It	Write It	Example
A **point** is an exact location in space.	A •	point *A*	point *A*	
A **line** is a straight path of points that continues without end in both directions.	B C	line *BC* line *CB*	$\overleftrightarrow{BC}$ $\overleftrightarrow{CB}$	
A **line segment** is part of a line between two endpoints.	D E	line segment *DE* line segment *ED*	$\overline{DE}$ $\overline{ED}$	YIELD
A **ray** is a part of a line that has one endpoint and continues without end in one direction.	F G	ray *FG*	$\overrightarrow{FG}$	ONE WAY →

🔑 Activity 1 Draw and label $\overline{JK}$.

_____

- Is there another way to name $\overline{JK}$? **Explain**.

_____

_____

MATHEMATICAL PRACTICES

Math Talk Explain how lines, line segments, and rays are related.

Angles

Term and Definition	Draw It	Read It	Write It	Example
An **angle** is formed by two rays or line segments that have the same endpoint. The shared endpoint is called the vertex.	*P* *Q* *R*	angle *PQR* angle *RQP* angle *Q*	∠*PQR* ∠*RQP* ∠*Q*	

You can name an angle by the vertex. When you name an angle using 3 points, the vertex is always the point in the middle.

Angles are classified by the size of the opening between the rays.

A **right angle** forms a square corner.	A **straight angle** forms a line.	An **acute angle** is less than a right angle.	An **obtuse angle** is greater than a right angle and less than a straight angle.

🔑 Activity 2 Classify an angle.

Materials ■ paper

To classify an angle, you can compare it to a right angle.

Make a right angle by using a sheet of paper. Fold the paper twice evenly to model a right angle. Use the right angle to classify the angles below.
Write *acute*, *obtuse*, *right*, or *straight*.

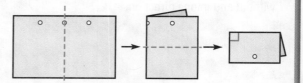

a.    b.    c.    d.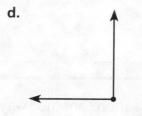

_____   _____   _____   _____

Name _____

Share and Show

1. Draw and label $\overline{AB}$ in the space at the right.

   $\overline{AB}$ is a _____ .

Draw and label an example of the figure.

2. $\overleftrightarrow{XY}$

3. obtuse ∠K

4. right ∠CDE

Use Figure M for 5 and 6.

5. Name a line segment.

6. Name a right angle.

Figure M

On Your Own

Draw and label an example of the figure.

7. $\overrightarrow{PQ}$

8. acute ∠RST

9. straight ∠WXZ

Use Figure F for 10–15.

10. Name a ray.

11. Name an obtuse angle.

_____

_____

12. Name a line.

13. Name a line segment.

_____

_____

14. Name a right angle.

15. Name an acute angle.

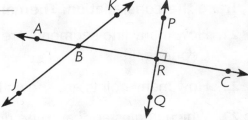

Figure F

Problem Solving REAL WORLD

Use the picture of the bridge for 16 and 17.

16. Classify ∠A.

17. Which angle appears to be obtuse?

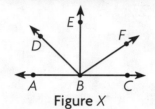

18. **H.O.T.** How many different angles are in Figure X? List them.

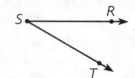

Figure X

19. **What's the Error?** Vanessa drew the angle at the right and named it ∠TRS. Explain why Vanessa's name for the angle is incorrect. Write a correct name for the angle.

20. **Test Prep** Which of the following terms best describes the figure at the right?

(A) ray   (C) line

(B) line segment   (D) angle

Connect to Science

Constellations

Astronomers study the stars and other objects in space. Cepheus is a constellation of stars named after an ancient mythological Greek king. Cepheus is visible in the northern sky all year long.

Trace the constellation. Then answer the questions.

21. How many line segments are shown in this drawing of Cepheus?

22. How many points are shown in this drawing of Cepheus?

23. Which angles appear to be right angles?

24. Which angle is an acute angle?

© Houghton Mifflin Harcourt Publishing Company

Name _____

 Mid-Chapter Checkpoint

▶ **Vocabulary**

Choose the best term from the box to complete the sentence.

1. A _____ is part of a line between two endpoints. (p. 381)

2. A _____ forms a square corner. (p. 382)

3. An _____ is greater than a right angle and less than a straight angle. (p. 382)

4. The two-dimensional figure that has one endpoint is a

_____. (p. 381)

5. An angle that forms a line is called a _____. (p. 382)

▶ **Concepts and Skills**

6. On the grid to the right, draw a polygon that has 2 pairs of parallel sides, 2 pairs of sides equal in length, and 2 acute and 2 obtuse angles. Tell all the possible names for the figure.

_____

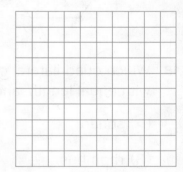

Draw the figure.

7. parallel lines	8. obtuse ∠ABC	9. intersecting lines that are not perpendicular	10. acute ∠RST

Fill in the bubble completely to show your answer.

11. Which statement is true?

   (A) A right triangle always has two acute angles.

   (B) An obtuse triangle always has two obtuse angles.

   (C) An acute triangle always has a right angle.

   (D) A triangle always has an obtuse angle.

12. Which figure has 2 pairs of sides that appear to be parallel?

   (A)   (C)

   (B)    (D)

13. Which quadrilateral can have 2 pairs of parallel sides, all sides with equal length, and no right angles?

   (A) square

   (B) rhombus

   (C) rectangle

   (D) trapezoid

14. Which names the figure correctly?

   (A) line *EF*

   (B) ray *FE*

   (C) angle *FE*

   (D) ray *EF*

Line Symmetry

Essential Question How can you check if a shape has line symmetry?

🔑 UNLOCK the Problem 〉REAL WORLD

One type of symmetry found in geometric shapes is line symmetry. This sign is in the hills above Hollywood, California. Do the letters in the Hollywood sign show line symmetry?

A shape has **line symmetry** if it can be folded about a line so that its two parts match exactly. A fold line, or a **line of symmetry**, divides a shape into two parts that are the same size and shape.

🔑 Activity Explore line symmetry.

Materials ■ pattern blocks ■ scissors ■ tracing paper

A Does the letter W have line symmetry?

STEP 1 Use pattern blocks to make the letter W.	**STEP 2** Trace the letter.

Math Idea

A vertical line goes up and down. ↕

A horizontal line goes left and right. ↔

A diagonal line goes through vertices of a polygon that are not next to each other. It can go up and down and left and right. ↗ ↙

STEP 3 Cut out the tracing.

STEP 4 Fold the tracing over a vertical line.

Think: The two parts of the folded W match exactly. The fold line is a line of symmetry.

Math Talk MATHEMATICAL PRACTICES
Why is it important to use a fold line to check if a shape has line symmetry?

So, the letter W _____ line symmetry.

B **Does the letter L have line symmetry?**

STEP 1

Use pattern blocks or grid paper to make the letter L.

STEP 2

Trace the letter.

STEP 3

Cut out the tracing.

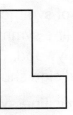

STEP 4

Fold the tracing over a vertical line.

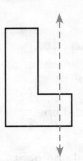

Do the two parts match exactly?

_____

STEP 5

Then open it and fold it horizontally.

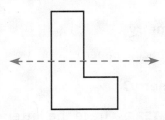

Do the two parts match exactly?

_____

STEP 6

Then open it and fold it diagonally.

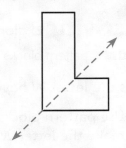

Do the two parts match exactly?

_____

So, the letter L _____ line symmetry.

1. Repeat Steps 1–6 for the remaining letters in HOLLYWOOD. Which letters have line symmetry?

   _____

2. Do any of the letters have more than one line of symmetry? **Explain.**

   _____

   _____

Remember

You can fold horizontally, vertically, or diagonally to determine if the parts match exactly.

Name _____

Share and Show

Tell whether the parts on each side of the line match.
Is the line a line of symmetry? Write *yes* or *no*.

1.

_____

2.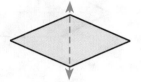

_____

3.

_____

☑ 4.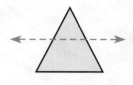

_____

Tell if the blue line appears to be a line of symmetry.
Write *yes* or *no*.

5.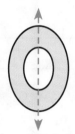

_____

6.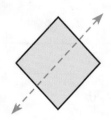

_____

7.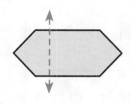

_____

☑ 8.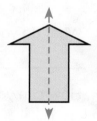

_____

Math Talk MATHEMATICAL PRACTICES
Explain how you can use paper folding to check if a shape has line symmetry.

On Your Own

Tell if the blue line appears to be a line of symmetry.
Write *yes* or *no*.

9.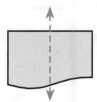

_____

10.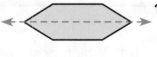

_____

11.

_____

12.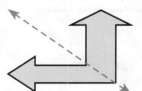

_____

H.O.T. Complete the design by reflecting over the line of symmetry.

13.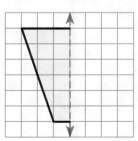

14.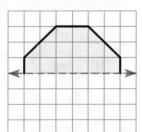

15.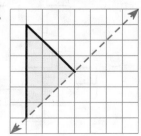

16.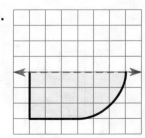

UNLOCK the Problem

17. Which shape has a correctly drawn line of symmetry?

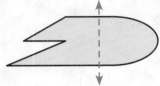

Ⓐ   Ⓒ

Ⓑ   Ⓓ

a. What do you need to find? _____

_____

b. How can you tell if the line of symmetry is correct?

_____

_____

c. Tell how you solved the problem.

_____

_____

d. Fill in the bubble for the correct answer choice above.

18. Which shape appears to have line symmetry?

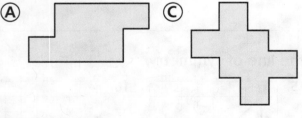

Ⓐ   Ⓒ

Ⓑ   Ⓓ

19. Which best describes the line of symmetry in the letter M?

Ⓐ horizontal

Ⓑ vertical

Ⓒ diagonal

Ⓓ rotational

FOR MORE PRACTICE:
Standards Practice Book, pp. P197–P198

Name _____

Find and Draw Lines of Symmetry

Essential Question How do you find lines of symmetry?

🔑 UNLOCK the Problem

How many lines of symmetry does each polygon have?

🔒 Activity 1 Find lines of symmetry.

Materials ■ isometric and square dot paper ■ straightedge

STEP 1

Draw a triangle like the one shown, so all sides have equal length.

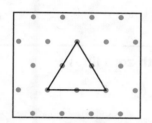

STEP 2

Fold the triangle in different ways to test for line symmetry. Draw along the fold lines that are lines of symmetry.

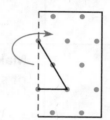

- Is there a line of symmetry if you fold the paper horizontally?

_____

STEP 3

Repeat the steps for each polygon shown. Complete the table.

Polygon	△ Triangle	◻ Square	▱ Parallelogram	◇ Rhombus	⬯ Trapezoid	⬡ Hexagon
Number of Sides	3					
Number of Lines of Symmetry	3					

- In a regular polygon, all sides are of equal length and all angles are equal. What do you notice about the number of lines of symmetry in regular polygons?

_____

_____

© Houghton Mifflin Harcourt Publishing Company

MATHEMATICAL PRACTICES

Math Talk How many lines of symmetry does a circle have? **Explain.**

○ Activity 2 Make designs that have line symmetry.

Materials ■ pattern blocks

Make a design by using more than one pattern block.
Record your design. Draw the line or lines of symmetry.

Make a design with 2 lines of symmetry.

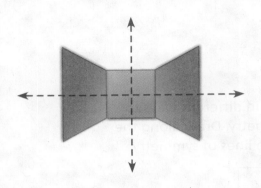

Make a design with 1 line of symmetry.

Make a design with more than 2 lines of symmetry.

Make a design with zero lines of symmetry.

Share and Show . . . .

1. The shape at the right has line symmetry.
   Draw the 2 lines of symmetry.

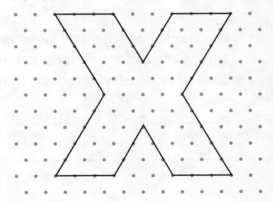

Name _____

Tell whether the shape appears to have zero lines, 1 line, or more than 1 line of symmetry. Write *zero, 1,* or *more than 1.*

2.
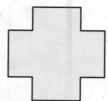

_____

☑ 3.

_____

4.

_____

☑ 5.

_____

Math Talk MATHEMATICAL PRACTICES
Explain how you can find lines of symmetry for a shape.

On Your Own

Tell whether the shape appears to have zero lines, 1 line, or more than 1 line of symmetry. Write *zero, 1,* or *more than 1.*

6.
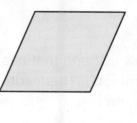

7.

8.

9.

Practice: Copy and Solve Does the design have line symmetry? Write *yes* or *no.* If your answer is *yes,* draw all lines of symmetry.

10.

11.
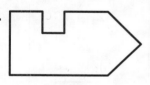

12.
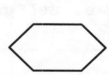

13.
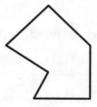

Draw a shape for each statement. Draw the line or lines of symmetry.

14. zero lines of symmetry

. . . . .
. . . . .
. . . . .
. . . . .
. . . . .

15. 1 line of symmetry

. . . . .
. . . . .
. . . . .
. . . . .
. . . . .

16. 2 lines of symmetry

. . . . .
. . . . .
. . . . .
. . . . .
. . . . .

Problem Solving

Use the chart for 17–19.

17. Which letters appear to have only 1 line of symmetry?

_____

18. Which letters appear to have zero lines of symmetry?

_____

19. The letter C has horizontal symmetry. The letter A has vertical symmetry. Which letters appear to have both horizontal and vertical symmetry?

_____

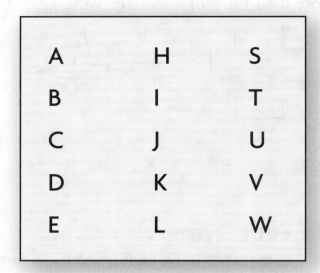

A	H	S
B	I	T
C	J	U
D	K	V
E	L	W

20.  **H.O.T.** **Sense or Nonsense?** Jeff says that the shape has only 2 lines of symmetry.

Does his statement make sense? **Explain.**

_____

_____

_____

_____

21. **Write Math** ▶ Draw a shape that has at least 2 lines of symmetry. Then write instructions that **explain** how to find the lines of symmetry.

_____

_____

_____

22. **Test Prep** How many lines of symmetry does the figure shown at the right have?

Ⓐ 0   Ⓒ 5

Ⓑ 1   Ⓓ 10

Name _____

Problem Solving • Shape Patterns

Essential Question How can you use the strategy *act it out* to solve pattern problems?

🔑 UNLOCK the Problem REAL WORLD

You can find patterns in fabric, pottery, rugs, and wall coverings. You can see patterns in shape, size, position, color, or number of figures.

Sofia will use the pattern below to make a wallpaper border. What might be the next three figures in the pattern?

Use the graphic organizer below to solve the problem.

Read the Problem

What do I need to find?	**What information do I need to use?**	**How will I use the information?**
I need to find the next three _____ in the pattern.	I need to use the _____ of each figure in Sofia's pattern.	I will use pattern blocks to model the _____ and act out the problem.

Solve the Problem

Describe how you acted out the problem to solve it.

I used a trapezoid and triangle to model the first

figure in the pattern. I used a _____ and

_____ to model the second figure in the pattern. I continued to model the pattern by repeating the models of the first two figures.

These are the next three figures in the pattern.

> **Math Talk** MATHEMATICAL PRACTICES
> **Explain** how you can describe the shape pattern using numbers.

🔑 Try Another Problem

Draw what might be the next figure in the pattern.

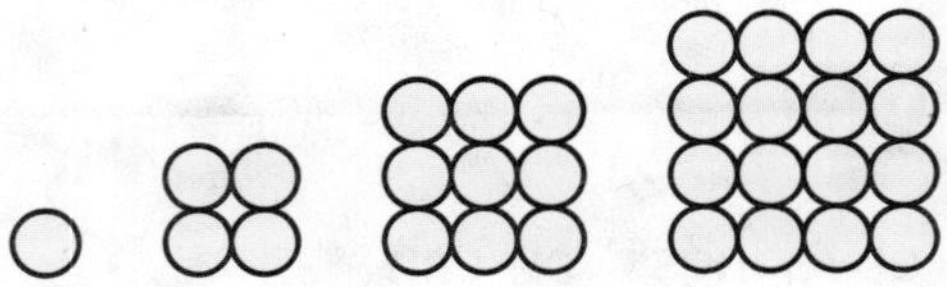

Figure: 1   2   3   4   _____
                                        5

How can you describe the pattern?

Read the Problem

What do I need to find?	What information do I need to use?	How will I use the information?

Solve the Problem

1. Use the figures to write a number pattern. Then describe the pattern in the numbers.

   _____

   _____

   _____

Math Talk MATHEMATICAL PRACTICES
What other strategy could you use to solve the problem?

2. What might the tenth number in your pattern be? **Explain.**

   _____

   _____

Name _____

Share and Show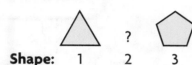

UNLOCK the Problem **Tips**

√ Use the Problem Solving MathBoard.
√ Underline the important facts.
√ Choose a strategy you know.

1. Marisol is making a pattern with blocks. What might the missing shape be?

First, look at the blocks.

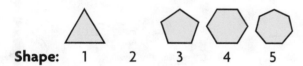

Shape: 1 2 3 4 5

Next, describe the pattern.

_____

_____

Finally, draw the missing shape.

Shape: 1 2 3 4 5

2. Use the shapes to write a number pattern. Then describe the pattern in the numbers.

_____

3. **H.O.T.** What if the pattern continued? Write an expression to describe the number of sides the sixth shape has in Marisol's pattern.

_____

4. Sahil made a pattern using circles. The first nine circles are shown. Describe the pattern. If Sahil continues the pattern, what might the next three circles be?

_____

© Houghton Mifflin Harcourt Publishing Company

On Your Own . . . .

Use the toy quilt designs for 5–6.

5. Lu is making a quilt that is 20 squares wide and has 24 rows. The border of the quilt is made by using each toy design equally as often. Each square can hold one design. How many of each design does she use for the border?

_____

6. **Write Math** ▶ Starting in the first square of her quilt, Lu lined up her toy designs in this order: plane, car, fire truck, helicopter, crane, and wagon. Using this pattern unit, which design will Lu place in the fifteenth square? Explain how you found your answer.

_____

_____

_____

_____

Choose a
STRATEGY

Act It Out
Draw a Diagram
Find a Pattern
Make a Table or List
Solve a Simpler Problem

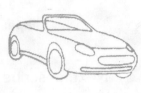
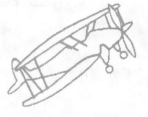

7. Missy uses 1 hexagonal, 2 rectangular, and 4 triangular pieces of fabric to make 1 bug design for a quilt. If she uses 70 pieces in all to make bug designs, how many of each shape does she use?

_____

SHOW YOUR WORK

8. **Test Prep** Neal has 3 square pattern blocks. How many lines of symmetry do all 3 pattern blocks have?

Ⓐ 3   Ⓑ 5   Ⓒ 6   Ⓓ 12

Name _____

Chapter Review/Test

▶ **Check Vocabulary**

Choose the best term from the box to complete the sentence.

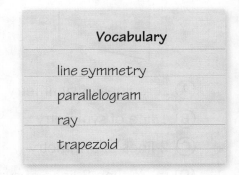
1. A _____ is a quadrilateral with exactly one pair of parallel sides. (p. 393)

2. A shape has _____ if it can be folded about a line so that its two parts match exactly. (p. 399)

3. A _____ has one endpoint and continues without end in one direction. (p. 381)

▶ **Check Concepts**

Tell if the blue line appears to be a line of symmetry.

Write _yes_ or _no_.

4.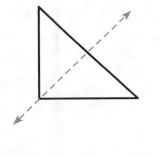
   _____

5.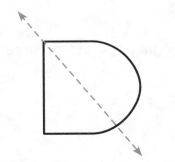
   _____

6.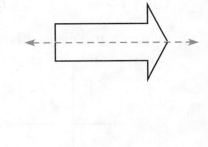
   _____

Use Figure _A_ for 7–9.

7. Name a pair of perpendicular lines.

   _____

8. Name a pair of intersecting lines that are not perpendicular.

   _____

9. Classify ∠AGD. Write _acute, right,_ or _obtuse._

   _____

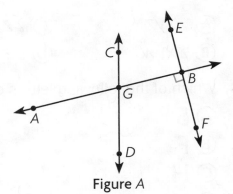

Figure _A_

GO Online Assessment Options
Chapter Test

Fill in the bubble completely to show your answer.

10. Which describes the shape?

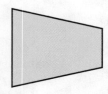

Ⓐ zero lines of symmetry

Ⓑ 1 line of symmetry

Ⓒ 2 lines of symmetry

Ⓓ more than 2 lines of symmetry

11. Which figure does **not** have two pairs of parallel sides?

Ⓐ parallelogram   Ⓒ rhombus

Ⓑ trapezoid   Ⓓ square

12. How many right angles can be in an obtuse triangle?

Ⓐ 0   Ⓒ 2

Ⓑ 1   Ⓓ 3

13. Which is the correct label for a right angle in the figure?

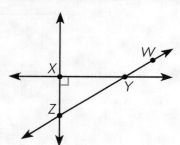

Ⓐ ∠XYZ   Ⓒ ∠ZXY

Ⓑ ∠XYW   Ⓓ ∠ZYX

14. Which of the following letters of the alphabet has line symmetry?

Ⓐ S

Ⓑ F

Ⓒ H

Ⓓ N

Name _____

Fill in the bubble completely to show your answer.

15. Which statement is true?

ⓐ A trapezoid can never have a right angle.

ⓑ A parallelogram can never have a right angle.

ⓒ A rhombus is a type of trapezoid.

ⓓ A square is a type of parallelogram.

16. Which lines appear parallel?

ⓐ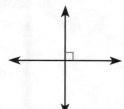

ⓒ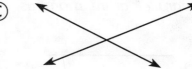

ⓑ

ⓓ

17. Norris drew the pattern below.

   _____ ?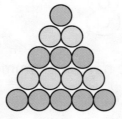

Which is the missing figure in the pattern?

ⓐ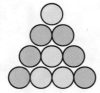

ⓒ

ⓑ

ⓓ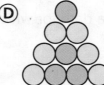

▶ Constructed Response

Describe a pattern. Write a rule using numbers to find the number of squares in any figure in the pattern.

18.

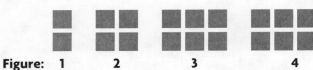

Figure: 1   2   3   4

Rule: _____

19. Classify the figure as many ways as possible. Write *quadrilateral, trapezoid, parallelogram, rhombus, rectangle,* or *square.*

_____

▶ Performance Task

20. Evie's birthday is the 18th day of May. Since May is the 5th month, Evie wrote the date like this:

5/18

Ⓐ Evie says all the numbers she wrote have line symmetry. Is she correct? Explain your thinking.

_____

_____

Ⓑ Choose one of the numbers Evie wrote. Using a straightedge, draw a line of symmetry.

Ⓒ Using the same format as Evie, write a date for which all the numbers have line symmetry.

414

11 Angles

Show What You Know ✓

Check your understanding of important skills.

Name _____

▶ **Use a Metric Ruler** Use a centimeter ruler to measure.
Find the length in centimeters.

1.

_____ centimeters

2.

_____ centimeters

▶ **Classify Angles** Classify the angle. Write *acute, right,* or *obtuse.*

3.

_____

4.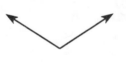

_____

5.

_____

▶ **Parts of a Whole** Write a fraction for each shaded part.

6. _____

7. _____

8. _____

9. _____

MATH DETECTIVE
WITH
CARMEN SANDIEGO™

The Sunshine Skyway Bridge crosses over Tampa Bay, Florida. Bridges and other building structures can model geometric figures. Be a Math Detective and investigate the bridge. Describe the geometric figures you see. Then classify the labeled angles and triangle.

Vocabulary Builder

▶ **Visualize It** •••
Complete the Bubble Map using review words.

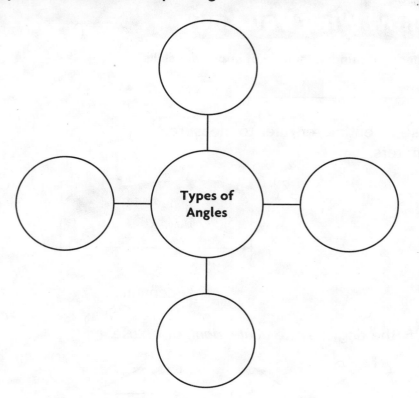

Review Words

acute

circle

obtuse

ray

right

straight

vertex

Preview Words

clockwise

counterclockwise

degree (°)

protractor

▶ **Understand Vocabulary** ••••••••••••••••••••••••••••••
Draw a line to match each word with its definition.

1. protractor

2. degree (°)

3. clockwise

4. counterclockwise

• In the same direction in which the hands of a clock move

• In the opposite direction in which the hands of a clock move

• A tool for measuring the size of an angle

• The unit used for measuring angles

GO Online • eStudent Edition • Multimedia eGlossary

Name _____

Angles and Fractional Parts of a Circle

Essential Question How can you relate angles and fractional parts of a circle?

Investigate

Materials ■ fraction circles

A. Place a $\frac{1}{12}$ piece on the circle. Place the tip of the fraction piece on the center of the circle. Trace the fraction piece.

What figure is formed by the fraction piece? _____

What parts of the fraction piece represent the rays of

the angle? _____

On what part of the circle is the vertex of the angle?

_____

B. Shade the angle formed by the $\frac{1}{12}$ piece. Label it $\frac{1}{12}$.

C. Place the $\frac{1}{12}$ piece back on the shaded angle. Turn it counterclockwise. **Counterclockwise** is the direction opposite from the way the hands move on a clock.

Trace the fraction piece in its new position. How many twelfths have you

traced in all? _____ Label $\frac{2}{12}$.

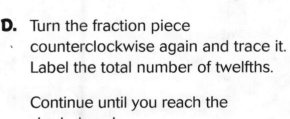

D. Turn the fraction piece counterclockwise again and trace it. Label the total number of twelfths.

Continue until you reach the shaded angle.

What figure is formed by turning and tracing the fraction piece? _____

How many times did you need to turn the $\frac{1}{12}$ piece to make a circle? _____

How many angles come together in the center of the circle? _____

Draw Conclusions

1. **Compare** the size of the angle formed by a $\frac{1}{4}$ piece and the size of the angle formed by a $\frac{1}{12}$ piece. Use a $\frac{1}{4}$ piece and your model on page 417 to help.

_____

_____

_____

2. **H.O.T.** **Synthesize** Describe the relationship between the size of the fraction piece and the number of turns it takes to make a circle.

_____

_____

Make Connections

You can relate fractions and angles to the hands of a clock.

Let the hands of the clock represent the rays of an angle. Each 5-minute mark represents a $\frac{1}{12}$ turn **clockwise**.

15 minutes elapse.

The minute hand makes a

_____ turn clockwise.

30 minutes elapse.

The minute hand makes a

_____ turn clockwise.

45 minutes elapse.

The minute hand makes a

_____ turn clockwise.

60 minutes elapse.

The minute hand makes a

_____ turn clockwise.

MATHEMATICAL PRACTICES

Math Talk Explain how an angle formed in a circle using a $\frac{1}{4}$ fraction piece is like a $\frac{1}{4}$ turn and 15 minutes elapsing on a clock.

Name _____

Share and Show

Tell what fraction of the circle the shaded angle represents.

1.

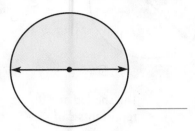

_____

2.

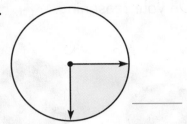

_____

3.

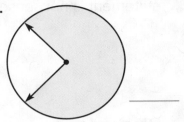

_____

4.

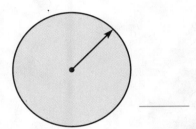

_____

5.

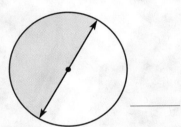

_____

6.

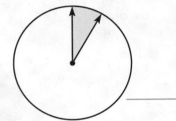

_____

Tell whether the angle on the circle shows a $\frac{1}{4}$, $\frac{1}{2}$, $\frac{3}{4}$, or 1 full turn clockwise or counterclockwise.

7.

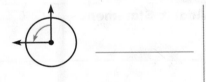

_____

_____

8.

_____

_____

9.

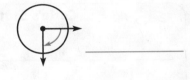

_____

_____

10.

_____

_____

11.

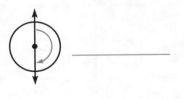

_____

_____

12.

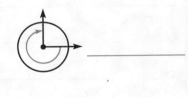

_____

_____

13. **H.O.T.** Susan watched the game from 1 P.M. to 1:30 P.M. **Describe** the turn the minute hand made.

_____

14. **Write Math** ▶ Compare the angles in Exercises 1 and 5. Does the position of the angle affect the size of the angle? **Explain.**

_____

_____

Problem Solving

 **Sense or Nonsense?**

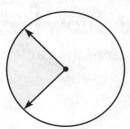

15. Whose statement makes sense? Whose statement is nonsense? **Explain** your reasoning.

The shaded angle represents $\frac{1}{4}$ of the circle.

The shaded angle represents $\frac{3}{8}$ of the circle.

Carla's Statement	**Adam's Statement**

• For the statement that is nonsense, write a statement that makes sense.

_____

• What is another way to describe the size of the angle? **Explain.**

_____

_____

Name _____

Degrees

Essential Question How are degrees related to fractional parts of a circle?

CONNECT You can use what you know about angles and fractional parts of a circle to understand angle measurement. Angles are measured in units called **degrees.** Think of a circle divided into 360 equal parts. An angle that turns through $\frac{1}{360}$ of the circle measures 1 degree.

> **Math Idea**
> The symbol for degrees is °.

 UNLOCK the Problem REAL WORLD

The angle between two spokes on the bicycle wheel turns through $\frac{10}{360}$ of a circle. What is the measure of the angle between the spokes?

• What part of an angle does a spoke represent?

_____

_____

🔑 **Example 1** Use fractional parts to find the angle measure.

Each $\frac{1}{360}$ turn measures _____ degree.

Ten $\frac{1}{360}$ turns measure _____ degrees.

So, the measure of the angle between the spokes is _____.

Math Talk MATHEMATICAL PRACTICES
How many degrees is the measure of an angle that turns through 1 whole circle? **Explain.**

▲ **The Penny Farthing bicycle was built in the 1800s.**

🔑 Example 2　Find the measure of a right angle.

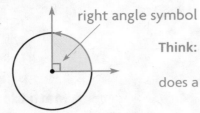

right angle symbol

Think: Through what fraction of a circle

does a right angle turn? _____

STEP 1 Write $\frac{1}{4}$ as an equivalent fraction with 360 in the denominator.

$$\frac{1}{4} = \frac{}{360}$$　Think: $4 \times 9 = 36$, so $4 \times$ _____ $= 360$.

> **Remember**
> To write an equivalent fraction, multiply the numerator and denominator by the same factor.

STEP 2 Write $\frac{90}{360}$ in degrees.

An angle that turns through $\frac{1}{360}$ of a circle measures _____.

An angle that turns through $\frac{90}{360}$ of a circle measures _____.

So, a right angle measures _____.

Try This! Find the measure of a straight angle.

Through what fraction of a circle does a straight angle turn? _____

Write $\frac{1}{2}$ as an equivalent fraction with 360 in the denominator.

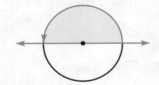

$$\frac{1}{2} = \frac{}{360}$$　Think: $2 \times 18 = 36$, so $2 \times$ _____ $= 360$.

So, a straight angle measures _____.

1. How can you describe the measure of an acute angle in degrees?

_____

_____

2. How can you describe the measure of an obtuse angle in degrees?

_____

Name _____

Share and Show

1. Find the measure of the angle.

   Through what fraction of a circle does the angle turn? _____

   $\dfrac{1}{3} = \dfrac{}{360}$ **Think:** $3 \times 12 = 36$, so $3 \times$ _____ $= 360$.

   So, the measure of the angle is _____.

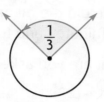

Tell the measure of the angle in degrees.

 2.

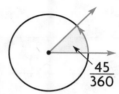

$\dfrac{45}{360}$

_____

3.

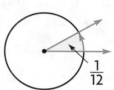

$\dfrac{1}{12}$

_____

On Your Own

Tell the measure of the angle in degrees.

Math Talk MATHEMATICAL PRACTICES If an angle measures 60°, through what fraction of a circle does it turn? **Explain.**

4.

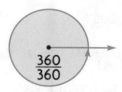

$\dfrac{360}{360}$

_____

5.

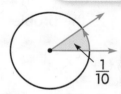

$\dfrac{1}{10}$

_____

Classify the angle. Write *acute, obtuse, right,* or *straight.*

6.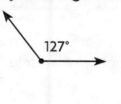

127°

_____

7.

_____

8.

37°

_____

9.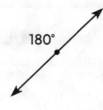

180°

_____

Classify the triangle. Write *acute, obtuse,* or *right.*

10.

45°

45°

_____

11.

40°

70° 70°

_____

12.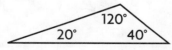

120°

20°   40°

_____

UNLOCK the Problem REAL WORLD

13. Ava started reading at 3:30 P.M. She stopped for a snack at 4:15 P.M. During this time, through what fraction of a circle did the minute hand turn? How many degrees did the minute hand turn?

a. What are you asked to find? _____

_____

b. What information can you use to find the fraction of a circle through which the minute hand turned?

_____

_____

c. How can you use the fraction of a circle through which the minute hand turned to find how many degrees it turned?

_____

_____

d. Show the steps to solve the problem.

STEP 1 $\dfrac{3 \times }{4 \times } = \dfrac{?}{360}$

_____

STEP 2 $\dfrac{3 \times 90}{4 \times 90} = \dfrac{}{360}$

e. Complete the sentences.

From 3:30 P.M. to 4:15 P.M., the minute hand made a _____ turn clockwise.

The minute hand turned _____ degrees.

---

14. ⚡H.O.T. **Write Math** ▶ Is this angle measure obtuse? **Explain.**

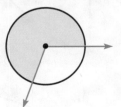

_____

_____

_____

15. Test Prep How many degrees are in an angle that turns through $\frac{2}{4}$ of a circle?

Ⓐ 90°

Ⓑ 180°

Ⓒ 270°

Ⓓ 360°

Measure and Draw Angles

Essential Question How can you use a protractor to measure and draw angles?

🔑 UNLOCK the Problem — REAL WORLD

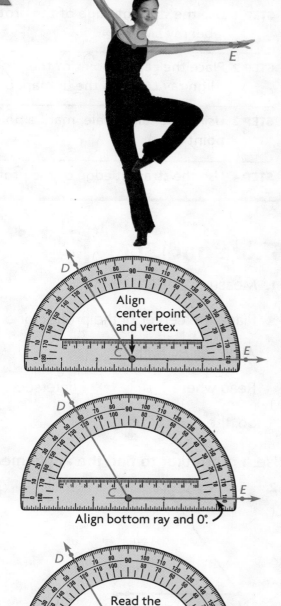

Emma wants to make a clay sculpture of her daughter as she appears in the photo from her dance recital. How can she measure ∠DCE, or the angle formed by her daughter's arms?

A **protractor** is a tool for measuring the size of an angle.

🔑 Activity Measure ∠DCE using a protractor.

Materials ■ protractor

STEP 1 Place the center point of the protractor on vertex C of the angle.

Align center point and vertex.

_____

STEP 2 Align the 0° mark on the scale of the protractor with ray CE.

Align bottom ray and 0°.

_____

STEP 3 Find where ray CD intersects the same scale. Read the angle measure on that scale. Extend the ray if you need to.

Read the scale.

The m∠DCE = _____.   Read the m∠DCE as the "measure of angle DCE".

So, the angle formed by Emma's daughter's

arms is _____.

MATHEMATICAL PRACTICES

Math Talk Can you line up either ray of the angle with the protractor when measuring? Explain.

Draw Angles You can also use a protractor to draw an angle of a given measure.

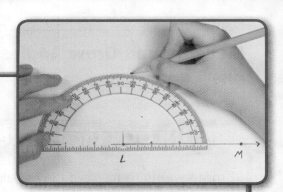

🔓 Activity Draw ∠KLM with a measure of 82°.

Materials ■ protractor

STEP 1 Use the straight edge of the protractor to draw and label ray *LM*.

STEP 2 Place the center point of the protractor on point *L*. Align ray *LM* with the 0° mark on the protractor.

STEP 3 Using the same scale, mark a point at 82°. Label the point *K*.

STEP 4 Use the straight edge of the protractor to draw ray *LK*.

Share and Show

1. Measure ∠ABC.

   Place the center of the protractor on point _____.

   Align ray *BC* with _____.

   Read where _____ intersects the same scale.

   So, the m∠ABC is _____.

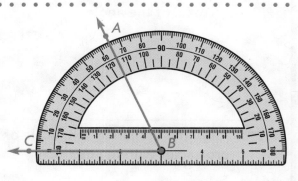

Use a protractor to find the angle measure.

2.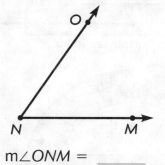

   m∠ONM = _____

✓ 3.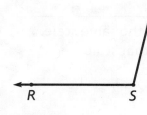

   m∠TSR = _____

⚠ **ERROR Alert**

Be sure to use the correct scale on the protractor. Ask yourself: Is the measure reasonable?

Use a protractor to draw the angle.

4. 170°

✓ 5. 78°

MATHEMATICAL PRACTICES

Math Talk Describe how drawing and measuring angles are similar.

© Houghton Mifflin Harcourt Publishing Company

426

Name _____

On Your Own ....

Use a protractor to find the angle measure.

6.

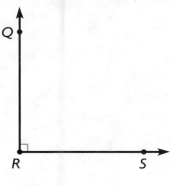

m∠QRS = _____

7.

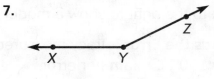

m∠XYZ = _____

Use a protractor to draw the angle.

8. 115°

9. 67°

Draw an example of each. Label the angle with its measure.

10. an acute angle

11. an obtuse angle

12. a straight angle

13. a right angle

14. **H.O.T.** **Write Math** ▶ Draw an angle with a measure of 0°.
Describe your drawing.

_____

_____

Problem Solving REAL WORLD

15. Mrs. Murphy is building a wheelchair ramp outside her business. The angle of the ramp should be 5°. Draw a picture in the space to the right to show a model of the ramp.

16. H.O.T. **What's the Error?** Tracy measured an angle as 50° that was actually 130°. **Explain** her error.

_____

_____

17. Test Prep What is the measure of ∠QRS?

Ⓐ 35°   Ⓑ 45°   Ⓒ 135°   Ⓓ 155°

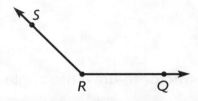

Connect to Science

Earth's Axis

Earth revolves around the sun yearly. The Northern Hemisphere is the half of Earth that is north of the equator. The seasons of the year are due to the tilt of Earth's axis.

Use the diagrams and a protractor for 18–20.

18. In the Northern Hemisphere, Earth's axis is tilted away from the sun on the first day of winter, which is often on December 21. What is the measure of the marked angle on the first day of winter, the shortest day of the year?

_____

19. Earth's axis is not tilted away from or toward the sun on the first days of spring and fall, which are often on March 20 and September 22. What is the measure of the marked angle on the first day of spring or fall?

_____

20. In the Northern Hemisphere, Earth's axis is tilted toward the sun on the first day of summer, which is often on June 21. What is the measure of the marked angle on the first day of summer, the longest day of the year?

_____

Northern Hemisphere

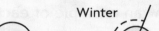
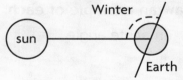

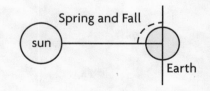

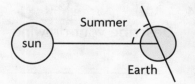

Name _____

 Mid-Chapter Checkpoint

▶ **Vocabulary**

Choose the best term from the box.

1. The unit used to measure an angle is called

   a _____. (p. 421)

2. _____ is the opposite of the
   direction in which the hands of a clock move. (p. 417)

3. A _____ is a tool for measuring the size
   of an angle. (p. 425)

▶ **Concepts and Skills**

Tell whether the angle on the circle shows a $\frac{1}{4}$, $\frac{1}{2}$, $\frac{3}{4}$, or 1 full
turn clockwise or counterclockwise.

4.

5.

6.

7.

_____

_____

_____

_____

Tell the measure of the angle in degrees.

8.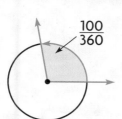

$\frac{100}{360}$

9.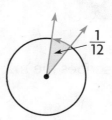

$\frac{1}{12}$

_____

_____

Use a protractor to draw the angle.

10. 75°

11. 127°

Fill in the bubble completely to show your answer.

12. Phillip watched a beach volleyball game from 1:45 P.M. to 2:00 P.M. How many degrees did the minute hand turn?

(A) 90°

(B) 180°

(C) 270°

(D) 360°

13. Which piece of pie forms a 180° angle?

(A)

(C)

(B)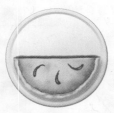

(D)

14. Which best describes the m∠CBT? Use a protractor to help you.

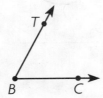

(A) acute; 58°

(C) obtuse; 62°

(B) acute; 62°

(D) obtuse; 118°

Name _____

Join and Separate Angles

Essential Question How can you determine the measure of an angle separated into parts?

Investigate

Materials ■ construction paper ■ scissors ■ protractor

A. Use construction paper. Draw an angle that measures exactly 70°. Label it ∠ABC.

B. Cut out ∠ABC.

C. Separate ∠ABC by cutting it into two parts. Begin cutting at the vertex and cut between the rays.

What figures did you form? _____

D. Use a protractor to measure the two angles you formed.

Record the measures. _____

E. Find the sum of the angles you formed.

_____ + _____ = _____
 part   + part   = whole

F. Join the two angles. Compare the m∠ABC to the sum of the measures of its parts. **Explain** how they compare.

_____

_____

_____

_____

> **Math Idea**
> You can think of ∠ABC as the whole and the two angles you formed as the parts of the whole.

Draw Conclusions

1. **What if** you cut ∠ABC into two different angles? What can you conclude about the sum of the measures of these two angles? **Explain.**

   _____

   _____

   _____

2. **H.O.T.** Seth cut ∠ABC into 3 parts. Draw a model that shows two different ways he could have separated his angle.

3. **Generalize** Write a sentence that compares the measure of an angle to the sum of its parts.

   _____

   _____

Make Connections

Materials ■ protractor

You can write the measure of the angles shown in a circle as a sum.

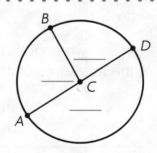

STEP 1 Use a protractor to find the measure of each angle.

_____

STEP 2 Label each angle with its measure.

_____

STEP 3 Write the sum of the angle measures as an equation.

$$\underline{\quad\quad} + \underline{\quad\quad} + \underline{\quad\quad} = \underline{\quad\quad}$$

part + part + part = whole

Math Talk Describe the angles shown in the circle above using the words *whole* and *part*.

Name _____

Share and Show

Add to find the measure of the angle. Write an
equation to record your work.

1.

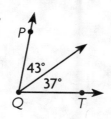

m∠PQT = _____

2.

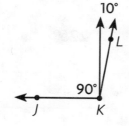

m∠JKL = _____

3.

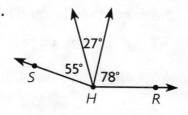

m∠RHS = _____

Use a protractor to find the measure of each angle.
Label each angle with its measure. Write the sum of
the angle measures as an equation.

4.

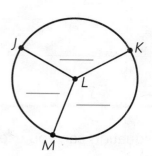

_____

5.

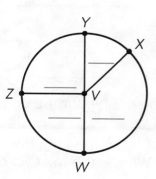

_____

6. **H.O.T.** **Write Math** ▶ Look back at Exercise 1. Suppose
you joined an angle measuring 10° to ∠PQT. **Describe** and
draw the new angle. Include all three parts in your drawing
and description.

_____

_____

_____

_____

🔑 UNLOCK the Problem REAL WORLD

7. Stephanie, Kay, and Shane each ate an equal-sized piece of a pizza. The measure of the angle of each piece was 45°. When the pieces were together, what is the measure of the angle they formed?

Ⓐ 90°   Ⓑ 135°   Ⓒ 180°   Ⓓ 225°

a. What are you asked to find? _____

_____

b. What information do you need to use? _____

_____

_____

c. Tell how you can use addition to solve the problem. _____

_____

_____

_____

d. Fill in the bubble for the correct answer choice above.

---

8. ⭐H.O.T.⭐ What is the m∠QRT?

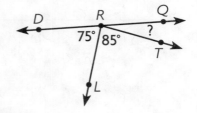

Ⓐ 20°

Ⓑ 150°

Ⓒ 160°

Ⓓ 180°

9. Which equation can you use to find the m∠XZW?

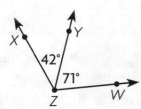

Ⓐ 71° − 42° = ■

Ⓑ 71° + 42° = ■

Ⓒ 71° × 42° = ■

Ⓓ 180° − 113° = ■

Name _____

Problem Solving • Unknown Angle Measures

Essential Question How can you use the strategy *draw a diagram* to solve angle measurement problems?

UNLOCK the Problem · REAL WORLD

Mr. Tran is cutting a piece of kitchen tile as shown at the right. He needs tiles with 45° angles to make a design. After the cut, what is the angle measure of the part left over? Can Mr. Tran use both pieces in the design?

Use the graphic organizer below to solve the problem.

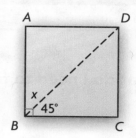

Read the Problem

What do I need to find?	What information do I need to use?	How will I use the information?
I need to find _____ _____	I can use the measures of the angles I know. _____ _____	I can draw a bar model and use the information to _____ _____

Solve the Problem

I can draw a bar model to represent the problem. Then I can write an equation to solve the problem.

45°	x

90°

$m\angle ABD + m\angle CBD = m\angle ABC$

$x + \underline{\hspace{1cm}} = \underline{\hspace{1cm}}$

$x = \underline{\hspace{1cm}}$

The $m\angle ABD = \underline{\hspace{1cm}}$.

Since both tiles measure _____, Mr. Tran can use both pieces in the design.

Math Talk MATHEMATICAL PRACTICES What other equation can you write to solve the problem? **Explain.**

Try Another Problem

Marisol is building a frame for a sandbox, but the boards she has are too short. She must join two boards together to build a side as shown. At what angle did she cut the first board?

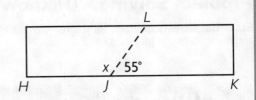

Read the Problem

What do I need to find?	What information do I need to use?	How will I use the information?

Solve the Problem

- **Explain** how you can check the answer to the problem.

_____

_____

Name _____

Share and Show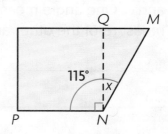

1. Laura cuts a square out of scrap paper as shown. What is the angle measure of the piece left over?

   First, draw a bar model to represent the problem.

   Next, write the equation you need to solve.

   _____

   _____

   Last, find the angle measure of the piece left over.
   The m∠MNQ = _____.
   So, the angle measure of the piece left over is _____.

2. **H.O.T.** **What if** Laura cut a smaller square as shown? Would the m∠MNQ be different? **Explain.**

   _____

   _____

   _____

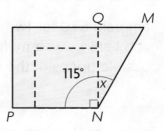

3. Jackie trimmed a piece of scrap metal to make a straight edge as shown. What is the measure of the piece she trimmed off?

   _____

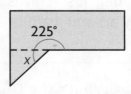

4. **H.O.T.** The map shows Marco's paper route. When Marco turns right onto Center Street from Main Street, what degree turn does he make? **Hint:** Draw a dashed line to extend Oak Street to form a 180° angle.

   _____

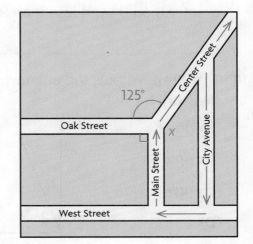

On Your Own . . . .

Choose a STRATEGY

Act It Out
Draw a Diagram
Find a Pattern
Make a Table or List
Solve a Simpler Problem

5. **Write Math** ▸ Two angles form a straight angle. One angle measures 89°. What is the measure of the other angle? **Explain.**

_____

_____

_____

6. **H.O.T.** **Pose a Problem** Look back at Problem 5. Write a similar problem about two angles that form a right angle.

_____

_____

_____

SHOW YOUR WORK

7. Sam paid $20 for two t-shirts. The price of each t-shirt was a multiple of 5. What are the possible prices of the t-shirts?

_____

8. Zayna has 3 boxes with 15 art books in each box. She has 2 bags with 11 math books in each bag. If she gives 30 books away, how many art and math books does she have now?

_____

9. **H.O.T.** **What's the Question?** It measures greater than 0° and less than 90°.

_____

10. **Test Prep** What is the unknown angle measure?

    (A) 22°

    (B) 68°

    (C) 90°

    (D) 158°

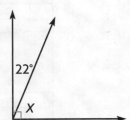

FOR MORE PRACTICE:
Standards Practice Book, pp. P215–P216

Chapter Review/Test

▶ Vocabulary

Choose the best term from the box.

1. The size of an angle can be measured using a tool called

   a _____. (p. 425)

2. _____ is the direction in which the hands of
   a clock move. (p. 418)

▶ Concepts and Skills

**Tell what fraction of the circle the shaded angle
represents.**

3.

4.

5.

_____   _____   _____

Use a protractor to draw the angle.

6. 68°

7. 145°

8. Use a protractor to find the measure of each angle.
   Label each angle with its measure. Write the sum
   of the angle measures as an equation.

   _____

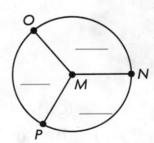

GO Online Assessment Options
Chapter Test

Fill in the bubble completely to show your answer.

9. Which describes the turn the angle on the circle shows?

Ⓐ 90° clockwise

Ⓑ 90° counterclockwise

Ⓒ 180° clockwise

Ⓓ 180° counterclockwise

10. Which best describes the m∠RST? Use a protractor to help you.

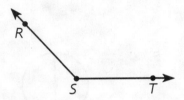

Ⓐ acute; 48°

Ⓑ obtuse; 48°

Ⓒ obtuse; 132°

Ⓓ obtuse; 148°

11. The pocket watch was invented in 1524. The time is 6 P.M. After 1 hour, how many degrees does the minute hand turn?

Ⓐ 45°　　　Ⓒ 180°

Ⓑ 90°　　　Ⓓ 360°

Fill in the bubble completely to show your answer.

12. What is the unknown angle measure?

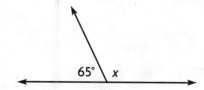

Ⓐ 25°

Ⓑ 115°

Ⓒ 125°

Ⓓ 180°

13. Which equation can you use to find the m∠WRT?

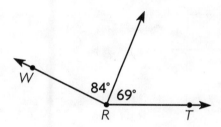

Ⓐ 84° + 69° = ■

Ⓑ 84° − 69° = ■

Ⓒ 84° × 69° = ■

Ⓓ 180° − 153° = ■

14. If an angle measures 100°, through what fraction of a circle does the angle turn?

Ⓐ $\frac{1}{100}$

Ⓑ $\frac{1}{4}$

Ⓒ $\frac{100}{360}$

Ⓓ $\frac{1}{2}$

▶ Constructed Response

15. How many right angles are there in an angle that turns through 360°? **Explain** how you know.

_____

_____

_____

_____

16. Soccer practice began at 2:30 P.M. and stopped at 3:00 P.M. because of rain. During this time, through what fraction of a circle did the minute hand turn? How many degrees did the minute hand turn? **Explain.**

_____

_____

_____

_____

▶ Performance Task

17. Charlotte divided a whole pizza into 4 pieces. One piece formed a straight angle. One piece formed a right angle. Two pieces formed acute angles with the same degree measure.

A Draw angles to represent the 4 pieces.

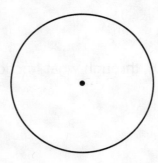

B Label each angle with its degree measure.

C Label each angle as a fraction of a circle.

D Write an equation that represents the degree measure of the whole pizza as the sum of the measures of its parts.

_____

© Houghton Mifflin Harcourt Publishing Company

Chapter 12

Relative Sizes of Measurement Units

Show What You Know ✓

Check your understanding of important skills.

Name _____

▶ **Time to the Half Hour** **Read the clock. Write the time.**

1.

_____

2.

_____

3.

_____

▶**Multiply by 1-Digit Numbers** **Find the product.**

4. 84
   × 7

5. 536
   × 8

6. 748
   × 5

7. 2,524
   ×   2

8. 360
   × 9

9. 296
   × 3

10. $1,428
   ×   4

11. 64
   × 5

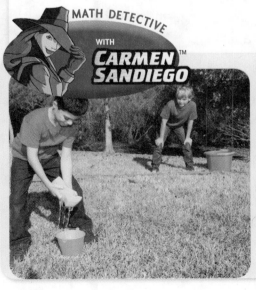

MATH DETECTIVE WITH **CARMEN SANDIEGO**™

A team was given a bucket of water and a sponge. The team had **1 minute** to fill an empty half-gallon bucket with water using only the sponge. The line plot shows the amount of water squeezed into the bucket. Be a Math Detective. Did the team squeeze enough water to fill the half-gallon bucket?

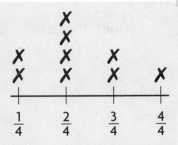

Amount of Water Squeezed into the Bucket (in cups)

Vocabulary Builder

▶ **Visualize It**••••••••••••••••••••••••••••••••••

Complete the Brain Storming diagram by using words with a ✓.

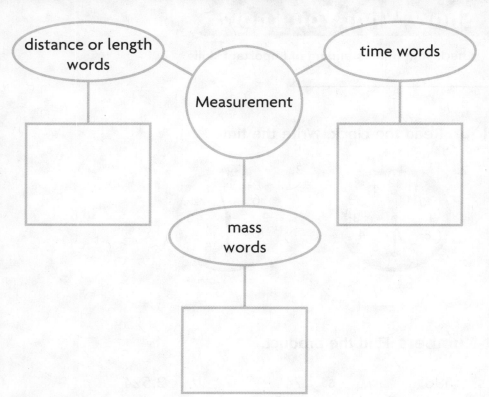

Review Words

✓ A.M.

✓ centimeter

✓ elapsed time

✓ foot

✓ gram

✓ hour

✓ inch

✓ kilogram

✓ meter

✓ minute

✓ P.M.

✓ yard

Preview Words

cup

decimeter

fluid ounce

gallon

half gallon

line plot

milliliter

millimeter

ounce

pint

pound

quart

second

ton

▶ **Understand Vocabulary**•••••••••••••••••••••••••

Draw a line to match each word with its definition.

1. decimeter

2. second

3. fluid ounce

4. ton

5. line plot

• A customary unit for measuring liquid volume

• A graph that shows the frequency of data along a number line

• A customary unit used to measure weight

• A small unit of time

• A metric unit for measuring length or distance

Name _____

Measurement Benchmarks

Essential Question How can you use benchmarks to understand the relative sizes of measurement units?

🔓 UNLOCK the Problem · REAL WORLD

Jake says the length of his bike is about four yards. Use the benchmark units below to determine if Jake's statement is reasonable.

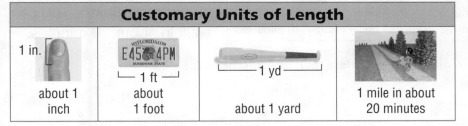

Customary Units of Length			
1 in. about 1 inch	E45 4PM └ 1 ft ┘ about 1 foot	└——— 1 yd ———┘ about 1 yard	1 mile in about 20 minutes

A **mile** is a customary unit for measuring length or distance. The benchmark shows the distance you can walk in about 20 minutes.

A baseball bat is about one yard long. Since Jake's bike is shorter than four times the length of a baseball bat, his bike is shorter than four yards long.

So, Jake's statement _____ reasonable.

Jake's bike is about _____ baseball bats long.

🔑 Example 1 Use the benchmark customary units.

Customary Units of Liquid Volume				
CUP 1 cup = 8 fluid ounces	1 pint	1 quart	1 half gallon	1 gallon

- About how much liquid is in a mug of hot chocolate? _____

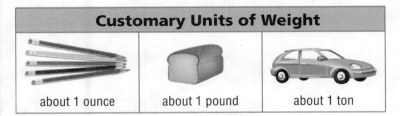

Customary Units of Weight		
about 1 ounce	about 1 pound	about 1 ton

- About how much does a grapefruit weigh? _____

Math Talk MATHEMATICAL PRACTICES
Order the units of weight from heaviest to lightest. Use benchmarks to **explain** your answer.

Benchmarks for Metric Units The metric system is based on place value. Each unit is 10 times as large as the next smaller unit. Below are some common metric benchmarks.

🔑 Example 2 Use the benchmark metric units.

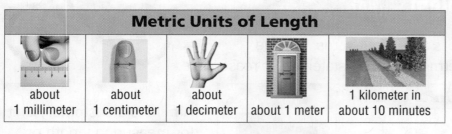

Metric Units of Length

| about 1 millimeter | about 1 centimeter | about 1 decimeter | about 1 meter | 1 kilometer in about 10 minutes |

A **kilometer** is a metric unit for measuring length or distance. The benchmark shows the distance you can walk in about 10 minutes.

• Is the length of your classroom greater than or less than one kilometer?

_____

Metric Units of Liquid Volume

| 1 milliliter | 1 liter |

• About how much medicine is usually in a medicine bottle?

about 120 _____

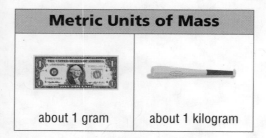

Metric Units of Mass

| about 1 gram | about 1 kilogram |

• About how much is the mass of a paper clip?

_____

© Houghton Mifflin Harcourt Publishing Company

Math Talk MATHEMATICAL PRACTICES
Explain how benchmark measurements can help you decide which unit to use when measuring.

Name _____

Share and Show

Use benchmarks to choose the metric unit you would use to measure each.

Metric Units
centimeter
meter
kilometer
gram
kilogram
milliliter
liter

1. mass of a strawberry

2. length of a cell phone

_____   _____

Circle the better estimate.

3. width of a teacher's desk

   10 meters or 1 meter

4. the amount of liquid a punch bowl holds

   2 liters or 20 liters

5. distance between Seattle and San Francisco

   6 miles or 680 miles

Math Talk MATHEMATICAL PRACTICES
Explain why you would use kilometers instead of meters to measure the distance across the United States.

On Your Own

Use benchmarks to choose the customary unit you would use to measure each.

Customary Units
inch
foot
yard
ounce
pound
cup
gallon

6. length of a football field

7. weight of a pumpkin

_____   _____

Circle the better estimate.

8. weight of a watermelon

   4 pounds or 4 ounces

9. the amount of liquid a fish tank holds

   10 cups or 10 gallons

Complete the sentence. Write *more* or *less*.

10. Matthew's large dog weighs _____ than one ton.

11. There can be _____ than one cup of water in a kitchen sink.

12. A paper clip has a mass of _____ than one kilogram.

Problem Solving REAL WORLD

Solve. For 13–15, use benchmarks to explain your answer.

13. Cristina is making macaroni and cheese for her family. Would Cristina use 1 pound of macaroni or 1 ounce of macaroni?

_____

_____

_____

14. Which is the better estimate for the length of a kitchen table, 200 centimeters or 200 meters?

_____

_____

_____

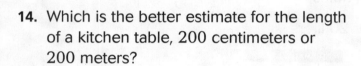

SHOW YOUR WORK

15. Amy thinks her dog weighs about 15 tons. Is this a reasonable estimate?

_____

_____

16. H.O.T. **Write Math** Dalton used benchmarks to estimate that there are more cups than quarts in one gallon. Is Dalton's estimate reasonable? **Explain.**

_____

_____

_____

17. Test Prep Which is the best estimate for a dose of medicine?

(A) 2 milliliters   (C) 2 millimeters

(B) 2 liters   (D) 2 meters

Name _____

Customary Units of Length

Essential Question How can you use models to compare customary units of length?

🔑 UNLOCK the Problem 〉 REAL WORLD

You can use a ruler to measure length. A ruler that is 1 foot long shows 12 inches in 1 foot. A ruler that is 3 feet long is called a yardstick. There are 3 feet in 1 yard.

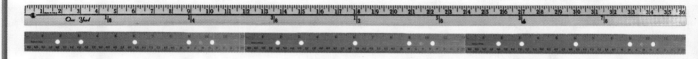

How does the size of a foot compare to the size of an inch?

🔑 Activity

Materials ■ 1-inch grid paper ■ scissors ■ tape

STEP 1 Cut out the paper inch tiles. Label each tile 1 inch.

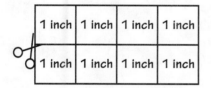

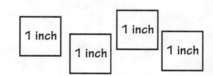

STEP 2 Place 12 tiles end-to-end to build 1 foot. Tape the tiles together.

							1 foot					
1 inch	1 inch	1 inch	1 inch	1 inch	1 inch	1 inch	1 inch	1 inch	1 inch	1 inch	1 inch

STEP 3 Compare the size of 1 foot to the size of 1 inch.

							1 foot					
1 inch	1 inch	1 inch	1 inch	1 inch	1 inch	1 inch	1 inch	1 inch	1 inch	1 inch	1 inch

Think: You need 12 inches to make 1 foot.

1 inch
1 inch

So, 1 foot is _____ times as long as 1 inch.

Math Talk MATHEMATICAL PRACTICES
How many inches would you need to make a yard? **Explain.**

🔑 Example Compare measures.

Emma has 4 feet of thread. She needs 50 inches of thread to make some bracelets. How can she determine if she has enough thread to make the bracelets?

Since 1 foot is 12 times as long as 1 inch, you can write feet as inches by multiplying the number of feet by 12.

STEP 1 Make a table that relates feet and inches.

Feet	Inches
1	12
2	
3	
4	
5	

Think:

1 foot × 12 = 12 inches

2 feet × 12 = _____

3 feet × _____ = _____

4 feet × _____ = _____

5 feet × _____ = _____

STEP 2 Compare 4 feet and 50 inches.

4 feet   50 inches

Think: Write each measure in inches and compare using <, >, or =.

_____ ◯ _____

Emma has 4 feet of thread. She needs 50 inches of thread.

4 feet is _____ than 50 inches.

So, Emma _____ enough thread to make the bracelets.

Math Talk MATHEMATICAL PRACTICES Explain how making a table helped you solve the problem.

- **What if** Emma had 5 feet of thread? Would she have enough thread to make the bracelets? **Explain.**

_____

_____

_____

Name _____

Share and Show

1. Compare the size of a yard to the size of a foot.
   Use a model to help.

Customary Units of Length
1 foot (ft) = 12 inches (in.)
1 yard (yd) = 3 feet
1 yard (yd) = 36 inches

1 yard

_____	_____	_____

1 yard is _____ times as long as _____ foot.

Complete.

2. 2 feet = _____ inches

3. 3 yards = _____ feet

4. 7 yards = _____ feet

> **Math Talk** MATHEMATICAL PRACTICES
> If you measured the length of your classroom in yards and then in feet, which unit would have a greater number of units? **Explain.**

On Your Own

Complete.

5. 4 yards = _____ feet

6. 10 yards = _____ feet

7. 7 feet = _____ inches

Algebra Compare using <, >, or =.

8. 1 foot ◯ 13 inches

9. 2 yards ◯ 6 feet

10. 6 feet ◯ 60 inches

Problem Solving REAL WORLD

11. **Write Math** ▶ Joanna has 3 yards of fabric. She needs 100 inches of fabric to make curtains. Does she have enough fabric to make curtains? **Explain.** Make a table to help.

Yards	Inches
1	
2	
3	

_____

_____

12. **Test Prep** Jim has 12 yards of carpet to cover his basement floor. He knows the length of his basement in feet. How many feet of carpet does he have?

Ⓐ 4 feet   Ⓒ 36 feet

Ⓑ 15 feet   Ⓓ 432 feet

H.O.T. **Sense or Nonsense?**

13. Jasmine and Luke used fraction strips to compare the size
of a foot to the size of an inch using fractions. They drew
models to show their answers. Whose answer makes sense?
Whose answer is nonsense? **Explain** your reasoning.

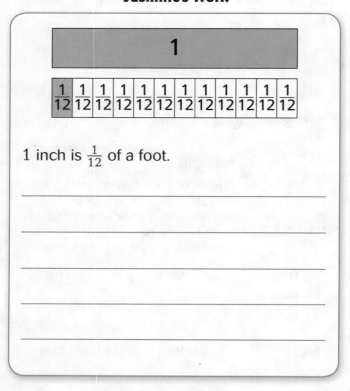

Jasmine's Work

1 inch is $\frac{1}{12}$ of a foot.

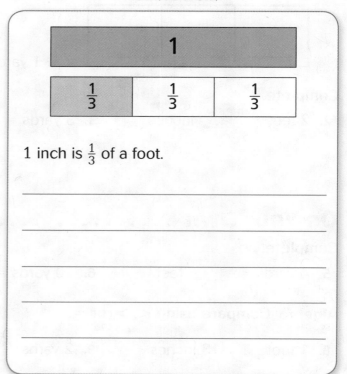

Luke's Work

1 inch is $\frac{1}{3}$ of a foot.

a. For the answer that is nonsense, write an answer that makes sense.

b. Look back at Luke's model. Which two units could you compare
using his model? **Explain.**

Name _____

Customary Units of Weight

Essential Question How can you use models to compare customary units of weight?

🔑 UNLOCK the Problem

Ounces and **pounds** are customary units of weight. How does the size of a pound compare to the size of an ounce?

🔑 Activity

Materials ■ color pencils

The number line below shows the relationship between pounds and ounces.

Pounds 0   1

Ounces 0 1 2 3 4 5 6 7 8 9 10 11 12 13 14 15 16

▲ You can use a spring scale to measure weight.

STEP 1 Use a color pencil to shade 1 pound on the number line.
_____
STEP 2 Use a different color pencil to shade 1 ounce on the number line.
_____
STEP 3 Compare the size of 1 pound to the size of 1 ounce.

You need _____ ounces to make _____ pound.

So, 1 pound is _____ times as heavy as 1 ounce.

MATHEMATICAL PRACTICES
Math Talk Which is greater, 9 pounds or 9 ounces? Explain.

• **Explain** how the number line helped you to compare the sizes of the units.

_____

_____

_____

_____

🔑 Example   Compare measures.

Nancy needs 5 pounds of flour to bake pies for a festival. She has 90 ounces of flour. How can she determine if she has enough flour to bake the pies?

STEP 1 Make a table that relates pounds and ounces.

Pounds	Ounces
1	16
2	
3	
4	
5	

Think:

1 pound × 16 = 16 ounces

2 pounds × 16 = _____

3 pounds × _____ = _____

4 pounds × _____ = _____

5 pounds × _____ = _____

STEP 2 Compare 90 ounces and 5 pounds.

90 ounces   5 pounds

Think: Write each measure in ounces and compare using <, >, or =.

_____ ◯ _____

Nancy has 90 ounces of flour. She needs 5 pounds of flour.

90 ounces is _____ than 5 pounds.

So, Nancy _____ enough flour to make the pies.

Try This! There are 2,000 pounds in 1 **ton**.
Make a table that relates tons and pounds.

Tons	Pounds
1	2,000
2	
3	

1 ton is _____ times as heavy as 1 pound.

454

Name _____

Share and Show

1. 4 tons = _____ pounds

Customary Units of Weight
1 pound (lb) = 16 ounces (oz)
1 ton (T) = 2,000 pounds

Think: 4 tons × _____ = _____

Complete.

✓ 2. 5 tons = _____ pounds

3. 6 pounds = _____ ounces

On Your Own

Math Talk MATHEMATICAL PRACTICES
What equation can you use to solve Exercise 4? Explain.

Complete.

✓ 4. 7 pounds = _____ ounces

5. 6 tons = _____ pounds

Algebra Compare using >, <, or =.

6. 1 pound ◯ 15 ounces

7. 2 tons ◯ 2 pounds

Problem Solving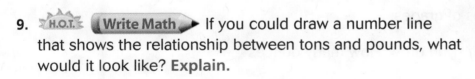

8. A landscaping company ordered 8 tons of gravel. They sell the gravel in 50 pound bags. How many pounds of gravel did the company order?

_____

9. **H.O.T.** **Write Math** ▶ If you could draw a number line that shows the relationship between tons and pounds, what would it look like? **Explain.**

_____

_____

10. **Test Prep** Kwadir is recording his baby sister's weight in pounds and in ounces each week. This week she weighs 10 pounds. How many ounces does she weigh?

(A) 10 ounces   (C) 20 ounces

(B) 16 ounces   (D) 160 ounces

H.O.T.   What's the Error?

11. Alexis bought $\frac{1}{2}$ pound of grapes. How many ounces of grapes did she buy?

Dan drew the number line below to solve the problem. He says his model shows that there are 5 ounces in $\frac{1}{2}$ pound. What is his error?

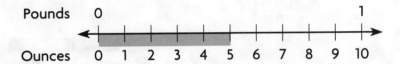

Look at the way Dan solved the problem. Find and describe his error.

Draw a correct number line and solve the problem.

So, Alexis bought _____ ounces of grapes.

- Look back at the number line you drew. How many ounces are in $\frac{1}{4}$ pound? **Explain.**

Customary Units of Liquid Volume

Essential Question How can you use models to compare customary units of liquid volume?

🔑 UNLOCK the Problem REAL WORLD

Liquid volume is the measure of the space a liquid occupies. Some basic units for measuring liquid volume are **gallons**, **half gallons**, **quarts**, **pints**, and **cups**.

The bars below model the relationships among some units of liquid volume. The largest units are gallons. The smallest units are **fluid ounces**.

| 1 cup = 8 fluid ounces |
| 1 pint = 2 cups |
| 1 quart = 4 cups |

1 gallon

1 gallon															
1 half gallon								1 half gallon							
1 quart				1 quart				1 quart				1 quart			
1 pint		1 pint		1 pint		1 pint		1 pint		1 pint		1 pint		1 pint	
1 cup	1 cup	1 cup	1 cup	1 cup	1 cup	1 cup	1 cup	1 cup	1 cup	1 cup	1 cup	1 cup	1 cup	1 cup	1 cup
8 fluid ounces	8 fluid ounces	8 fluid ounces	8 fluid ounces	8 fluid ounces	8 fluid ounces	8 fluid ounces	8 fluid ounces	8 fluid ounces	8 fluid ounces	8 fluid ounces	8 fluid ounces	8 fluid ounces	8 fluid ounces	8 fluid ounces	8 fluid ounces

🔒 Example How does the size of a gallon compare to the size of a quart?

MATHEMATICAL PRACTICES

Math Talk Describe the pattern in the units of liquid volume.

STEP 1 Draw two bars that represent this relationship. One bar should show gallons and the other bar should show quarts.

STEP 2 Shade 1 gallon on one bar and shade 1 quart on the other bar.

STEP 3 Compare the size of 1 gallon to the size of 1 quart.

So, 1 gallon is _____ times as much as 1 quart.

Example Compare measures.

Serena needs to make 3 gallons of lemonade for the lemonade sale. She has a powder mix that makes 350 fluid ounces of lemonade. How can she decide if she has enough powder mix?

STEP 1 Use the model on page 457. Find the relationship between gallons and fluid ounces.

1 gallon = _____ cups

1 cup = _____ fluid ounces

1 gallon = _____ cups × _____ fluid ounces

1 gallon = _____ fluid ounces

STEP 2 Make a table that relates gallons and fluid ounces.

Gallons	Fluid Ounces
1	128
2	
3	

Think:

1 gallon = 128 fluid ounces

2 gallons × 128 = _____ fluid ounces

3 gallons × 128 = _____ fluid ounces

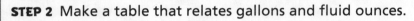

STEP 3 Compare 350 fluid ounces and 3 gallons.

350 fluid ounces   3 gallons

Think: Write each measure in fluid ounces and compare using <, >, or =.

_____   _____

Serena has enough mix to make 350 fluid ounces. She needs to make 3 gallons of lemonade.

350 fluid ounces is _____ than 3 gallons.

So, Serena _____ enough mix to make 3 gallons of lemonade.

458

Name _____

Share and Show

1. Compare the size of a quart to the size of a pint.
   Use a model to help.

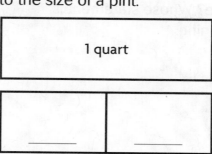

1 quart

_____   _____

Customary Units of Liquid Volume
1 cup (c) = 8 fluid ounces (fl oz)
1 pint (pt) = 2 cups
1 quart (qt) = 2 pints
1 quart (qt) = 4 cups
1 gallon (gal) = 4 quarts
1 gallon (gal) = 8 pints
1 gallon (gal) = 16 cups

1 quart is _____ times as much as _____ pint.

Complete.

✓ 2. 2 pints = _____ cups   3. 3 gallons = _____ quarts   ✓ 4. 6 quarts = _____ cups

On Your Own

MATHEMATICAL PRACTICES

Math Talk Explain how the conversion chart above relates to the bar model in Exercise 1.

Complete.

5. 4 gallons = _____ pints   6. 5 cups = _____ fluid ounces

Algebra Compare using >, <, or =.

7. 2 gallons ◯ 32 cups   8. 4 pints ◯ 6 cups   9. 5 quarts ◯ 11 pints

Problem Solving REAL WORLD

10. H.O.T. A soccer team has 25 players. The team's thermos holds 4 gallons of water. If the thermos is full, is there enough water for each player to have 2 cups? **Explain.** Make a table to help.

_____

_____

Gallons	Cups
1	
2	
3	
4	

11. **Test Prep** A pitcher contains 5 quarts of water. How many cups of water does the pitcher contain?

Ⓐ 4 cups   Ⓒ 20 cups

Ⓑ 10 cups   Ⓓ 40 cups

Problem Solving REAL WORLD

H.O.T. Sense or Nonsense?

12. Whose statement makes sense? Whose statement is nonsense? **Explain** your reasoning.

1 pint is $\frac{1}{4}$ of a gallon.

1 pint is $\frac{1}{8}$ of a gallon.

Zach's Statement	**Angela's Statement**

a. For the statement that is nonsense, write a statement that makes sense.

b. Describe the size of a pint as it relates to a quart using fractions.

Line Plots

Essential Question How can you make and interpret line plots with fractional data?

🔑 UNLOCK the Problem REAL WORLD

The data show the lengths of the buttons in Jen's collection. For an art project, she wants to know how many buttons are longer than $\frac{1}{4}$ inch.

You can use a line plot to solve the problem. A **line plot** is a graph that shows the frequency of data along a number line.

Length of Buttons in Jen's Collection (in inches)
$\frac{1}{4}, \frac{3}{4}, \frac{1}{4}, \frac{4}{4}, \frac{1}{4}, \frac{4}{4}$

Make a line plot to show the data.

🔑 Example 1

STEP 1 Order the data from least to greatest length and complete the tally table.

_____

_____

STEP 2 Label the fraction lengths on the number line below from the least value of the data to the greatest.

_____

STEP 3 Plot an *X* above the number line for each data point. Write a title for the line plot.

Buttons in Jen's Collection	
Length (in inches)	**Tally**
$\frac{1}{4}$	
$\frac{3}{4}$	
$\frac{4}{4}$	

_____   _____   _____   _____

_____

So, _____ buttons are longer than $\frac{1}{4}$ inch.

Math Talk MATHEMATICAL PRACTICES
Explain how you labeled the numbers on the number line in Step 2.

1. How many buttons are in Jen's collection? _____

2. What is the difference in length between the longest button and the shortest button in Jen's collection? _____

Think: To find the difference, subtract the numerators. The denominators stay the same.

🔑 Example 2

Some of the students in Ms. Lee's class walk to school. The data show the distances these students walk. What distance do most students walk?

Distance Students Walk to School (in miles)
$\frac{1}{2}, \frac{1}{2}, \frac{1}{4}, \frac{3}{4}, \frac{1}{4}, \frac{1}{2}, \frac{1}{2}$

Make a line plot to show the data.

STEP 1 Order the data from least to greatest distance and complete the tally table.

_____

STEP 2 Label the fraction lengths on the number line below from the least value of the data to the greatest.

STEP 3 Plot an X above the number line for each data point. Write a title for the line plot.

Distance Students Walk to School	
Distance (in miles)	Tally

So, most students walk _____.

3. How many more students walk $\frac{1}{2}$ mile than $\frac{1}{4}$ mile to school?

_____

4. What is the difference between the longest distance and the shortest distance that students walk?

_____

5. **What if** a new student joins Ms. Lee's class who walks $\frac{3}{4}$ mile to school? How would the line plot change? **Explain.**

_____

Name _____

Share and Show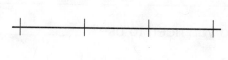

1. A food critic collected data on the lengths of time customers waited for their food. Order the data from least to greatest time. Make a tally table and a line plot to show the data.

Time Customers Waited for Food (in hours)
$\frac{1}{2}$, $\frac{1}{4}$, $\frac{3}{4}$, $\frac{1}{4}$, $\frac{1}{4}$, $\frac{1}{2}$, 1

Time Customers Waited for Food	
Time (in hours)	Tally

Math Talk Explain how the line plot helped you answer the question for Exercise 2.

MATHEMATICAL PRACTICES

Use your line plot for 2 and 3.

2. On how many customers did the food critic collect data? _____

3. What is the difference between the longest time and the shortest time that customers waited? _____

On Your Own

4. The data show the lengths of the ribbons Mia used to wrap packages. Make a tally table and a line plot to show the data.

Ribbon Length Used to Wrap Packages (in yards)
$\frac{1}{6}$, $\frac{2}{6}$, $\frac{5}{6}$, $\frac{3}{6}$, $\frac{2}{6}$, $\frac{6}{6}$, $\frac{3}{6}$, $\frac{2}{6}$

Ribbon Used to Wrap Packages	
Length (in yards)	Tally

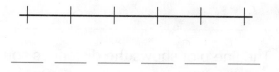

Use your line plot for 5.

5. What is the difference in length between the longest ribbon and the shortest ribbon Mia used? _____

🔑 UNLOCK the Problem REAL WORLD

6. The line plot shows the distances the students in Mr. Boren's class ran at the track in miles. Altogether, did the students run more or less than 5 miles?

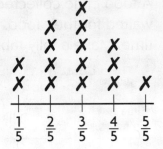

a. What are you asked to find? _____

_____

b. What information do you need to use? _____

_____

c. How will the line plot help you solve the problem? _____

_____

d. What operation will you use to solve the problem? _____

e. Show the steps to solve the problem.

_____

_____

f. Complete the sentences.

The students ran a total of _____ miles.

_____ miles _____ 5 miles; so, altogether

the students ran _____ than 5 miles.

---

7. 🔥 H.O.T. **Write Math** ▸ Lena collects antique spoons. The line plot shows the lengths of the spoons in her collection. If she lines up all of her spoons in order of size, what is the size of the middle spoon? **Explain.**

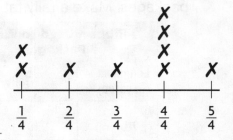

Length of Spoons (in feet)

_____

_____

8. Test Prep The line plot shows the distances some students hiked. What is the difference between the longest distance and the shortest distance the students hiked?

Ⓐ $\frac{1}{8}$ mile Ⓒ $\frac{7}{8}$ mile

Ⓑ $\frac{3}{8}$ mile Ⓓ $\frac{11}{8}$ mile

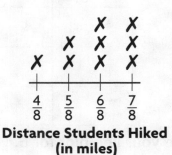

Distance Students Hiked (in miles)

© Houghton Mifflin Harcourt Publishing Company

Name _____

 Mid-Chapter Checkpoint

▶ **Vocabulary**

Choose the best term from the box to complete the sentence.

1. A _____ is a customary unit used to measure weight.
   (p. 453)

2. The cup and the _____ are both customary units for measuring liquid volume. (p. 457)

▶ **Concepts and Skills**

Complete the sentence. Write *more* or *less*.

3. A cat weighs _____ than one ounce.

4. Serena's shoe is _____ than one yard long.

Complete.

5. 5 feet = _____ inches

6. 4 tons = _____ pounds

7. 4 cups = _____ pints

8. Mrs. Byrne's class went raspberry picking. The data show the weights of the cartons of raspberries the students picked. Make a tally table and a line plot to show the data.

Weight of Cartons of Raspberries Picked (in pounds)
$\frac{3}{4}, \frac{1}{4}, \frac{2}{4}, \frac{4}{4}, \frac{1}{4}, \frac{1}{4}, \frac{2}{4}, \frac{3}{4}, \frac{3}{4}$

Cartons of Raspberries Picked	
Weight (in pounds)	**Tally**

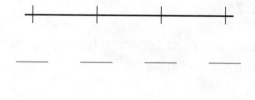

Use your line plot for 9 and 10.

9. What is the difference in weight between the heaviest carton

   and lightest carton of raspberries? _____

10. How many pounds of raspberries did Mrs. Byrne's class pick in all? _____

Fill in the bubble completely to show your answer.

11. A jug contains 2 gallons of water. How many quarts of water does the jug contain?

Ⓐ 4 quarts

Ⓑ 8 quarts

Ⓒ 16 quarts

Ⓓ 32 quarts

12. Serena bought 4 pounds of dough to make pizzas. The recipe gives the amount of dough needed for a pizza in ounces. How many ounces of dough did she buy?

Ⓐ 8 ounces

Ⓑ 16 ounces

Ⓒ 64 ounces

Ⓓ 96 ounces

13. Vaughn threw the shot put 9 yards at a track meet. The official used a tape measure to measure the distance in feet. How many feet did he throw the shot put?

Ⓐ 27 feet

Ⓑ 30 feet

Ⓒ 108 feet

Ⓓ 324 feet

14. What is the best estimate for the amount of liquid a watering can holds?

Ⓐ 5 ounces

Ⓑ 5 cups

Ⓒ 5 quarts

Ⓓ 5 gallons

Name _____

Metric Units of Length

Essential Question How can you use models to compare metric units of length?

Investigate

Materials ■ ruler (meter) ■ scissors ■ tape

Meters (m), **decimeters** (dm), centimeters (cm), and **millimeters** (mm) are all metric units of length.

Build a meterstick to show how these units are related.

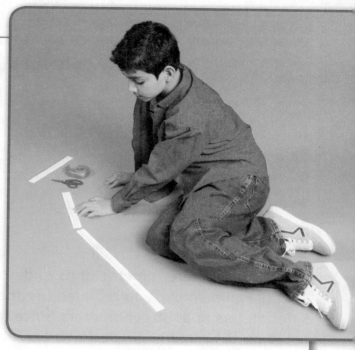

A. Cut out the meterstick strips.

B. Place the strips end-to-end to build 1 meter. Tape the strips together.

C. Look at your meter strip. What patterns do you notice about the sizes of the units?

1 meter is _____ times as long as 1 decimeter.

1 decimeter is _____ times as long as 1 centimeter.

1 centimeter is _____ times as long as 1 millimeter.

Describe the pattern you see.

_____

_____

> **Math Idea**
> If you lined up 1,000 metersticks end-to-end, the length of the metersticks would be 1 kilometer.

Draw Conclusions . . . .

1. **Compare** the size of 1 meter to the size of 1 centimeter. Use your meterstick to help.

_____

2. Compare the size of 1 meter to the size of 1 millimeter. Use your meterstick to help.

_____

_____

3. **H.O.T.** **Apply** What operation could you use to find how many centimeters are in 3 meters? **Explain.**

_____

_____

_____

_____

Make Connections . . . .

You can use different metric units to describe the same metric length. For example, you can measure the length of a book as 3 decimeters or as 30 centimeters. Since the metric system is based on the number 10, decimals or fractions can be used to describe metric lengths as equivalent units.

Think of 1 meter as one whole. Use your meterstick to write equivalent units as fractions and decimals.

1 meter = 10 decimeters	1 meter = 100 centimeters
Each decimeter is	Each centimeter is
_____ or _____ of a meter.	_____ or _____ of a meter.

Complete the sentence.

- A length of 51 centimeters is _____ or _____ of a meter.

- A length of 8 decimeters is _____ or _____ of a meter.

- A length of 82 centimeters is _____ or _____ of a meter.

Math Talk MATHEMATICAL PRACTICES
Explain how you are able to locate and write decimeters and centimeters as parts of a meter on the meterstick.

Name _____

Share and Show   . . . .

Complete.

Metric Units of Length
1 centimeter (cm) = 10 millimeters (mm)
1 decimeter (dm) = 10 centimeters
1 meter (m) = 10 decimeters
1 meter (m) = 100 centimeters
1 meter (m) = 1,000 millimeters

✅ **1.** 2 meters = _____ centimeters

2. 3 centimeters = _____ millimeters

3. 5 decimeters = _____ centimeters

Algebra Compare using <, >, or =.

4. 4 meters ◯ 40 decimeters

5. 5 centimeters ◯ 5 millimeters

6. 6 decimeters ◯ 65 centimeters

7. 7 meters ◯ 700 millimeters

**Describe the length in meters. Write your answer
as a fraction and as a decimal.**

✅ **8.** 65 centimeters = _____ or _____ meter

9. 47 centimeters = _____ or _____ meter

10. 9 decimeters = _____ or _____ meter

11. 2 decimeters = _____ or _____ meter

Problem Solving REAL WORLD

12. Lucille runs the 50-meter dash in her track meet.
How many decimeters long is the race?

_____

13. 🔆 H.O.T. Alexis is knitting a blanket 2 meters long. Every
2 decimeters, she changes the color of the yarn to make
stripes. How many stripes will the blanket have? **Explain.**

_____

_____

14. Write Math ▸ **Explain** how you know that a line that is 8 centimeters long is longer than a line that is 75 millimeters long.

_____

_____

_____

15. H.O.T. **What's the Error?** Julianne's desk is 75 centimeters long. She says her desk is 7.5 meters long. **Describe** her error.

_____

_____

_____

H.O.T. **Pose a Problem**

16. Aruna was writing a report on pecan trees. She made the table of information to the right.

Write a problem that can be solved by using the data.

Pecan Tree	
Average Measurements	
Length of nuts	3 cm to 5 cm
Height	21 m to 30 m
Width of trunk	18 dm
Width of leaf	10 cm to 20 cm

Pose a problem.

Solve your problem.

- **Describe** how you could change the problem by changing a unit in the problem. Then solve the problem.

_____

_____

FOR MORE PRACTICE:
Standards Practice Book, pp. P231–P232

Metric Units of Mass and Liquid Volume

Essential Question How can you use models to compare metric units of mass and liquid volume?

🔓 UNLOCK the Problem REAL WORLD

Mass is the amount of matter in an object. Metric units of mass include kilograms (kg) and grams (g). Liters (L) and **milliliters** (mL) are metric units of liquid volume.

The charts show the relationship between these units.

Metric Units of Mass
1 kilogram (kg) = 1,000 grams (g)

Metric Units of Liquid Volume
1 liter (L) = 1,000 milliliters (mL)

🔑 Example 1 Compare kilograms and grams.

Becky planted a flower garden full of bluebonnets. She used 9 kilograms of soil. How many grams of soil is that?

number of kilograms   grams in 1 kilogram   total grams

$$9 \quad \times \quad 1{,}000 \quad = \quad \rule{2cm}{0.4pt}$$

So, Becky used _____ grams of soil to plant her bluebonnets.

- Are kilograms larger or smaller than grams?

_____

- Will the number of grams be greater than or less than the number of kilograms?

_____

- What operation will you use to solve the problem?

_____

🔑 Example 2 Compare liters and milliliters.

Becky used 5 liters of water to water her bluebonnet garden. How many milliliters of water is that?

number of liters   milliliters in 1 liter   total milliliters

$$5 \quad \times \quad 1{,}000 \quad = \quad \rule{2cm}{0.4pt}$$

So, Becky used _____ milliliters of water.

Math Talk MATHEMATICAL PRACTICES
Compare the size of a kilogram to the size of a gram. Then compare the size of a liter to the size of a milliliter.

Share and Show

1. There are 3 liters of water in a pitcher. How many milliliters of water are in the pitcher?

There are _____ milliliters in 1 liter. Since I am changing

from a larger unit to a smaller unit, I can _____ 3 by 1,000 to find the number of milliliters in 3 liters.

So, there are _____ milliliters of water in the pitcher.

Complete.

2. 4 liters = _____ milliliters

3. 6 kilograms = _____ grams

Math Talk MATHEMATICAL PRACTICES
Explain how you found the number of grams in 6 kilograms in Exercise 3.

On Your Own

Complete.

4. 8 kilograms = _____ grams

5. 7 liters = _____ milliliters

Algebra Compare using <, >, or =.

6. 1 kilogram ◯ 900 grams

7. 2 liters ◯ 2,000 milliliters

Algebra Complete.

8.

Liters	Milliliters
1	1,000
2	
3	
	4,000
5	
6	
	7,000
8	
9	
10	

9.

Kilograms	Grams
1	1,000
2	
	3,000
4	
5	
6	
7	
	8,000
9	
10	

Name _____

Problem Solving REAL WORLD

10. Frank wants to fill a fish tank with 8 liters of water. How many milliliters is that?

_____

11. Kim has 3 water bottles. She fills each bottle with 1 liter of water. How many milliliters of water does she have?

_____

12. Jared's empty backpack has a mass of 3 kilograms. He doesn't want to carry more than 7 kilograms on a trip. How many grams of equipment can Jared pack?

_____

SHOW YOUR WORK

13. A large cooler contains 20 liters of iced tea and a small cooler contains 5 liters of iced tea. How many more milliliters of iced tea does the large cooler contain than the small cooler?

_____

14. **H.O.T.** A 500-gram bag of granola costs $4, and a 2-kilogram bag of granola costs $15. What is the cheapest way to buy 2,000 grams of granola? **Explain.**

_____

_____

_____

15. **Sense or Nonsense?** The world's largest apple had a mass of 1,849 grams. Sue said the mass was greater than 2 kilograms. Does Sue's statement make sense? **Explain.**

_____

_____

_____

UNLOCK the Problem   REAL WORLD

16. Lori bought 600 grams of cayenne pepper and 2 kilograms of black pepper. How many grams of pepper did she buy?

black pepper   cayenne pepper

a. What are you asked to find?

_____

b. What information will you use?

_____

c. Tell how you might solve the problem.

_____

_____

_____

d. Show how you solved the problem.

_____

_____

_____

_____

e. Complete the sentences.

Lori bought _____ grams of cayenne pepper.

She bought _____ grams of black pepper.

_____ + _____ = _____ grams

So, Lori bought _____ grams of pepper in all.

---

17. ▸ Write Math ▸ Jill has two rocks. One has a mass of 20 grams and the other has a mass of 20 kilograms. Which rock has the greater mass? **Explain.**

_____

_____

_____

18. **Test Prep** Caroline bought a bag of onions that was labeled 5 kilograms. She needs to know how many grams that is for her recipe. How many grams is 5 kilograms?

Ⓐ   50 grams

Ⓑ   500 grams

Ⓒ   5,000 grams

Ⓓ 50,000 grams

Units of Time

Essential Question How can you use models to compare units of time?

UNLOCK the Problem REAL WORLD

The analog clock below has an hour hand, a minute hand, and a **second** hand to measure time. The time is 4:30:12.

> **Read Math**
>
> Read 4:30:12 as 4:30 and 12 seconds, or 30 minutes and 12 seconds after 4.

• Are there more minutes or seconds in one hour?

_____

There are 60 seconds in a minute and 60 minutes in an hour. The clocks below show the length of a second, a minute, and an hour.

Start Time: 3:00:00

1 second elapses.

The time is now 3:00:01.

1 minute, or 60 seconds, elapses. The second hand has made a full turn clockwise.

The time is now 3:01:00.

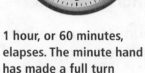

1 hour, or 60 minutes, elapses. The minute hand has made a full turn clockwise.

The time is now 4:00:00.

Example 1 How does the size of an hour compare to the size of a second?

There are _____ minutes in an hour.

There are _____ seconds in a minute.

60 minutes × _____ = _____ seconds

There are _____ seconds in a hour.

So, 1 hour is _____ times as long as 1 second.

Think: Multiply the number of minutes in a hour by the number of seconds in a minute.

Math Talk MATHEMATICAL PRACTICES How many full turns clockwise does a minute hand make in 3 hours? **Explain.**

🔑 Example 2 Compare measures.

Larissa spent 2 hours on her science project.
Cliff spent 200 minutes on his science project.
Who spent more time?

STEP 1 Make a table that relates hours and minutes.

Hours	Minutes
1	60
2	
3	

STEP 2 Compare 2 hours and 200 minutes.

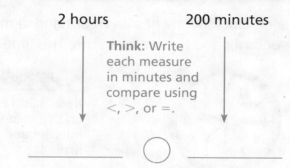

2 hours   200 minutes

Think: Write each measure in minutes and compare using $<$, $>$, or $=$.

_____ ◯ _____

2 hours is _____ than 200 minutes.

So, _____ spent more time than _____ on the science project.

🔑 Activity Compare the length of a week to the length of a day.

Materials ■ color pencils

The number line below shows the relationship between days and weeks.

STEP 1 Use a color pencil to shade 1 week on the number line.

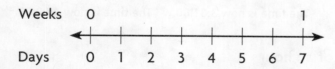

Weeks   0   1

Days   0   1   2   3   4   5   6   7

_____

STEP 2 Use a different color pencil to shade 1 day on the number line.

_____

STEP 3 Compare the size of 1 week to the size of 1 day.

There are _____ days in _____ week.

So, 1 week is _____ times as long as 1 day.

476

Name _____

Share and Show . . . .

1. Compare the length of a year to the length of a month. Use a model to help.

Years   0   1

Months   0 1 2 3 4 5 6 7 8 9 10 11 12

Math Talk MATHEMATICAL PRACTICES
Explain how the number line helped you compare the length of a year and the length of a month.

1 year is _____ times as long as _____ month.

Complete.

✓ 2. 2 minutes = _____ seconds

✓ 3. 4 years = _____ months

On Your Own . . . .

Complete.

4. 3 minutes = _____ seconds

5. 4 hours = _____ minutes

Algebra Compare using >, <, or =.

6. 3 years ◯ 35 months

7. 2 days ◯ 40 hours

Problem Solving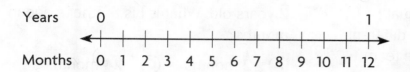

8. Damien has lived in the apartment building for 5 years. Ken has lived there for 250 weeks. Who has lived in the building longer? **Explain.** Make a table to help.

_____

_____

9. H.O.T. How many hours are in a week? **Explain.**

_____

_____

Years	Weeks
1	
2	
3	
4	
5	

10. **Write Math** ▶ **Explain** how you know that 9 minutes is less than 600 seconds.

_____

_____

11. **H.O.T.** Football practice lasts 3 hours. The coach wants to spend an equal number of minutes on each of 4 different plays. How many minutes will the team spend on each play?

_____

12. **Test Prep** Martin's brother just turned 2 years old. What is his brother's age in months?

(A) 2 months   (C) 24 months

(B) 14 months (D) 104 months

Connect to Science

One day is the length of time it takes Earth to make one complete rotation. One year is the time it takes Earth to revolve around the sun. To make the calendar match Earth's orbit time, there are leap years. Leap years add one extra day to the year. A leap day, February 29, is added to the calendar every four years.

| 1 year = 365 days |
| 1 leap year = 366 days |

13. How many days are there in 4 years, if the fourth year is a leap year? **Explain.** Make a table to help.

_____

_____

Years	Days
1	
2	
3	
4	

14. Parker was born on February 29, 2008. The second time he is able to celebrate on his actual birthday is in 2016. How many days old will Parker be on February 29, 2016? **Explain.**

_____

_____

_____

_____

Name _____

Problem Solving • Elapsed Time

Essential Question How can you use the strategy *draw a diagram* to solve elapsed time problems?

UNLOCK the Problem REAL WORLD

Dora and her brother Kyle spent 1 hour and 35 minutes doing yard work. Then they stopped for lunch at 1:20 P.M. At what time did they start doing yard work?

Use the graphic organizer to help you solve the problem.

Read the Problem

What do I need to find?	**What information do I need to use?**	**How will I use the information?**
I need to find the time that Dora and Kyle _____.	I need to use the _____ and the time that they _____.	I can draw a time line to help me count backward and find the _____.

Solve the Problem

I draw a time line that shows the end time 1:20 P.M. Next, I count backward 1 hour and then 5 minutes at a time until I have 35 minutes.

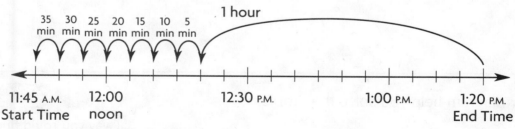

So, Dora and her brother Kyle started doing yard work at _____.

1. **What if** Dora and Kyle spent 50 minutes doing yard work and they stopped for lunch at 12:30 P.M.? What time would they have started doing yard work?

_____

🔑 Try Another Problem

Ben started riding his bike at 10:05 A.M. He stopped 23 minutes later when his friend Robbie asked him to play kickball. At what time did Ben stop riding his bike?

Read the Problem

What do I need to find?	What information do I need to use?	How will I use the information?

Solve the Problem

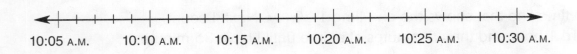

10:05 A.M.   10:10 A.M.   10:15 A.M.   10:20 A.M.   10:25 A.M.   10:30 A.M.

---

2. How did your diagram help you solve the problem?

_____

_____

_____

MATHEMATICAL PRACTICES

Math Talk Describe another way you could find the time an activity started or ended given the elapsed time and either the start or end time.

Name _____

Share and Show

♀ UNLOCK the Problem Tips

✓ Use the Problem Solving MathBoard.
✓ Choose a strategy you know.
✓ Underline important facts.

1. Evelyn has dance class every Saturday. It lasts
   1 hour and 15 minutes and is over at 12:45 P.M.
   At what time does Evelyn's dance class begin?

   First, write the problem you need to solve.

   _____

   Next, draw a time line to show the end time and
   the elapsed time.

11:00 A.M.   12:00   1:00 P.M.
                    noon

   Finally, find the start time.

   Evelyn's dance class begins at _____ .

2. **H.O.T.** **What if** Evelyn's dance class started
   at 11:00 A.M. and lasted 1 hour and 25 minutes?
   At what time would her class end? **Describe** how
   this problem is different from Problem 1.

   _____

   _____

3. Beth got on the bus at 8:06 A.M.
   Thirty-five minutes later, she arrived
   at school. At what time did Beth arrive
   at school?

   _____

   _____

4. Lyle went fishing for 1 hour and
   30 minutes until he ran out of bait
   at 6:40 P.M. At what time did Lyle
   start fishing?

   _____

   _____

On Your Own . . . .

Choose a
STRATEGY

Act It Out
Draw a Diagram
Find a Pattern
Make a Table or List
Solve a Simpler Problem

5. Mike and Jed went skiing at 10:30 A.M. They skied for 1 hour and 55 minutes before stopping for lunch. At what time did Mike and Jed stop for lunch?

_____

6. _H.O.T._ **What's the Question?** One hour and 10 minutes later, it was 6:20 P.M.

_____

_____

_____

SHOW YOUR WORK

7. _Write Math_ ► **Explain** how you can use a diagram to determine the start time when the end time is 9:00 A.M. and the elapsed time is 26 minutes. What is the start time?

_____

_____

_____

_____

_____

8. _H.O.T._ Bethany finished her math homework at 4:20 P.M. She did 25 multiplication problems in all. If each problem took her 3 minutes to do, at what time did Bethany start her math homework?

_____

9. Test Prep Vincent began his weekly chores on Saturday morning at 11:20. He finished 1 hour and 15 minutes later. At what time did Vincent finish his chores?

  Ⓐ 12:35 A.M.   Ⓒ 12:35 P.M.

  Ⓑ 10:05 A.M.   Ⓓ 1:05 P.M.

Mixed Measures

Essential Question How can you solve problems involving mixed measures?

🔑 UNLOCK the Problem REAL WORLD

Herman is building a picnic table for a new campground. The picnic table is 5 feet 10 inches long. How long is the picnic table in inches?

🔒 **Change a mixed measure.**

Think of 5 feet 10 inches as 5 feet + 10 inches.

Write feet as inches.

5 feet   **Think:** 5 feet × 12 = ⟶ [ ] inches
+ 10 inches   60 inches   + [ ] inches
                                    [ ] inches

So, the picnic table is _____ inches long.

- Is the mixed measure greater than or less than 6 feet?

_____

- How many inches are in 1 foot?

_____

🔑 Example 1 Add mixed measures.

Herman built the picnic table in 2 days. The first day he worked for 3 hours 45 minutes. The second day he worked for 2 hours 10 minutes. How long did it take him to build the table?

STEP 1 Add the minutes.

```
  3 hr 45 min
+ 2 hr 10 min
       [ ] min
```

STEP 2 Add the hours.

```
  3 hr 45 min
+ 2 hr 10 min
[ ] hr 55 min
```

So, it took Herman _____ to build the table.

Math Talk MATHEMATICAL PRACTICES How is adding mixed measures similar to adding tens and ones? How is it different? **Explain.**

- **What if** Herman worked an extra 5 minutes on the picnic table? How long would he have worked on the table then? **Explain.**

_____

_____

© Houghton Mifflin Harcourt Publishing Company

🔑 Example 2 Subtract mixed measures.

Alicia is building a fence around the picnic area. She has a pole that is 6 feet 6 inches long. She cuts off 1 foot 7 inches from one end. How long is the pole now?

STEP 1 Subtract the inches.

Think: 7 inches is greater than 6 inches. You need to regroup to subtract.

6 ft 6 in. = 5 ft 6 in. + 12 in.

           = 5 ft _____ in.

$$\begin{array}{r} \overset{5}{\cancel{6}}\text{ft}\ \overset{18}{\cancel{6}}\text{ in.} \\ -\ 1\text{ ft } 7\text{ in.} \\ \hline \boxed{}\text{ in.} \end{array}$$

 ERROR Alert
Be sure to check that you are regrouping correctly. There are 12 inches in 1 foot.

STEP 2 Subtract the feet.

$$\begin{array}{r} \overset{5}{\cancel{6}}\text{ft}\ \overset{18}{\cancel{6}}\text{ in.} \\ -\ 1\text{ ft } 7\text{ in.} \\ \hline \boxed{}\text{ ft } 11\text{ in.} \end{array}$$

So, the pole is now _____ long.

Try This! Subtract.

3 pounds 5 ounces − 1 pound 2 ounces

Share and Show

1. A truck is carrying 2 tons 500 pounds of steel. How many pounds of steel is the truck carrying?

Think of 2 tons 500 pounds as 2 tons + 500 pounds.
Write tons as pounds.

    2 tons   Think: 2 tons × 2,000 = ⟶   pounds

+ 500 pounds   _____ pounds   + _____ pounds

                                                                  pounds

So, the truck is carrying _____ pounds of steel.

Name _____

Rewrite each measure in the given unit.

2. 1 yard 2 feet

_____ feet

3. 3 pints 1 cup

_____ cups

✓ 4. 3 weeks 1 day

_____ days

Add or subtract.

5. 2 lb 4 oz
    + 1 lb 6 oz
    ‾‾‾‾‾‾‾‾‾‾

✓ 6. 3 gal 4 qt
     − 1 gal 5 qt
     ‾‾‾‾‾‾‾‾‾‾

7. 5 hr 20 min
    − 3 hr 15 min
    ‾‾‾‾‾‾‾‾‾‾‾

Math Talk MATHEMATICAL PRACTICES How do you know when you need to regroup to subtract? Explain.

On Your Own . . . .

Rewrite each measure in the given unit.

8. 1 hour 15 minutes

_____ minutes

9. 4 quarts 2 pints

_____ pints

10. 10 feet 10 inches

_____ inches

Add or subtract.

11. 2 tons 300 lb
     − 1 ton 300 lb
     ‾‾‾‾‾‾‾‾‾‾‾

12. 10 gal 8 c
     + 8 gal 9 c
     ‾‾‾‾‾‾‾‾‾

13. 7 lb 6 oz
     − 2 lb 12 oz
     ‾‾‾‾‾‾‾‾‾‾

Problem Solving REAL WORLD

14. H.O.T. Jackson has a rope 1 foot 8 inches long. He cuts it into 4 equal pieces. How many inches long is each piece?

_____

15. H.O.T. Ahmed fills 6 pitchers with juice. Each pitcher contains 2 quarts 1 pint. How many pints of juice does he have?

_____

16. H.O.T. **Sense or Nonsense?** Sam and Dave each solve the problem at the right. Sam says the sum is 4 feet 18 inches. Dave says the sum is 5 feet 6 inches. Whose answer makes sense? Whose answer is nonsense? **Explain.**

   2 ft 10 in.
 + 2 ft 8 in.
 ‾‾‾‾‾‾‾‾‾‾

_____

_____

🔑 UNLOCK the Problem > REAL WORLD

17. Theo is practicing for a 5-kilometer race. He runs
5 kilometers every day and records his time. His normal
time is 25 minutes 15 seconds. Yesterday it took him only
23 minutes 49 seconds. How much faster was his time
yesterday than his normal time?

Ⓐ 1 minute 26 seconds   Ⓒ 2 minutes 26 seconds

Ⓑ 1 minute 64 seconds   Ⓓ 2 minutes 34 seconds

a. What are you asked to find?

_____

b. What information do you know?

_____

_____

c. How will you solve the problem?

_____

d. Solve the problem.

e. Fill in the bubble for the correct answer
choice above.

---

18. H.O.T. Don has 5 pieces of pipe. Each
piece is 3 feet 6 inches long. If Don joins
the pieces end to end to make one long
pipe, how long will the new pipe be?

Ⓐ 8 feet 11 inches

Ⓑ 15 feet 6 inches

Ⓒ 15 feet 11 inches

Ⓓ 17 feet 6 inches

19. Maya's cat weighed 7 pounds 2 ounces
last year. The cat gained 1 pound 8
ounces this year. What is the weight of
Maya's cat now?

Ⓐ 5 pounds 10 ounces

Ⓑ 8 pounds 2 ounces

Ⓒ 8 pounds 10 ounces

Ⓓ 9 pounds

FOR MORE PRACTICE:
Standards Practice Book, pp. P239–P240

Patterns in Measurement Units

Essential Question How can you use patterns to write number pairs for measurement units?

CONNECT The table at the right relates yards and feet. You can think of the numbers in the table as number pairs. 1 and 3, 2 and 6, 3 and 9, 4 and 12, and 5 and 15 are number pairs.

The number pairs show the relationship between yards and feet. 1 yard is equal to 3 feet, 2 yards is equal to 6 feet, 3 yards is equal to 9 feet, and so on.

Yards	Feet
1	3
2	6
3	9
4	12
5	15

UNLOCK the Problem REAL WORLD

Lillian made the table below to relate two units of time. What units of time does the pattern in the table show?

Activity
Use the relationship between the number pairs to label the columns of the table.

___	___
1	7
2	14
3	21
4	28
5	35

- List the number pairs.

  _____

- **Describe** the relationship between the numbers in each pair.

  _____

  _____

- Label the columns of the table.   **Think:** What unit of time is 7 times as great as another unit?

MATHEMATICAL PRACTICES

Math Talk Look at each number pair in the table. Could you change the order of the numbers in the number pairs? **Explain** why or why not.

Try This! Jasper made the table below to relate two customary units of liquid volume. What customary units of liquid volume does the pattern in the table show?

- List the number pairs.

  _____

- **Describe** the relationship between the numbers in each pair.

  _____

  _____

  _____

- Label the columns of the table.

Think: What customary unit of liquid volume is 4 times as great as another unit?

1	4
2	8
3	12
4	16
5	20

- What other units could you have used to label the columns of the table above? **Explain.**

  _____

  _____

  _____

Share and Show MATH BOARD ·

1. The table shows a pattern for two units of time. Label the columns of the table with the units of time.

   Think: What unit of time is 24 times as great as another unit?

1	24
2	48
3	72
4	96
5	120

Math Talk MATHEMATICAL PRACTICES
Explain how you labeled the columns of the table.

Name _____

Each table shows a pattern for two customary units. Label
the columns of the table.

2.

_____	_____
1	2
2	4
3	6
4	8
5	10

3.

_____	_____
1	16
2	32
3	48
4	64
5	80

On Your Own

Each table shows a pattern for two units of time. Label
the columns of the table.

4.

_____	_____
1	60
2	120
3	180
4	240
5	300

5.

_____	_____
1	12
2	24
3	36
4	48
5	60

Each table shows a pattern for two metric units of length.
Label the columns of the table.

6.

_____	_____
1	10
2	20
3	30
4	40
5	50

7.

_____	_____
1	100
2	200
3	300
4	400
5	500

8. **Write Math** List the number pairs for the table in Exercise 6.
Describe the relationship between the numbers in each pair.

_____

_____

_____

Problem Solving REAL WORLD

9. **H.O.T.** **What's the Error?** Maria wrote *Weeks* as the label for the first column of the table and *Years* as the label for the second column. **Describe** her error.

?	?
1	52
2	104
3	156
4	208
5	260

_____

_____

_____

10. **H.O.T.** **Sense or Nonsense?** The table shows a pattern for two metric units. Lou labels the columns *Meters* and *Millimeters*. Zayna labels them *Liters* and *Milliliters*. Whose answer makes sense? Whose answer is nonsense? **Explain.**

?	?
1	1,000
2	2,000
3	3,000
4	4,000
5	5,000

_____

_____

_____

_____

11. Look back at Problem 10. What other labels for metric units could you write for the columns of the table? **Explain.**

_____

_____

12. **H.O.T.** Look at the following number pairs: 1 and 365, 2 and 730, 3 and 1,095. The number pairs describe the relationship between which two units of time? **Explain.**

_____

_____

13. **Test Prep** The table shows a pattern for two customary units of length. Which are the best labels?

(A) Years, Months   (C) Yards, Inches

(B) Feet, Inches   (D) Yards, Feet

?	?
1	12
2	24
3	36
4	48
5	60

Name _____

 **Chapter Review/Test**

▶ **Vocabulary**

Choose the best term from the box to complete the sentence.

Vocabulary
gram
line plot
milliliter
millimeter
quart

1. A _____ is a metric unit for measuring length or distance. (p. 467)

2. A _____ is a metric unit for measuring liquid volume. (p. 471)

3. A _____ is a graph that shows the frequency of data along a number line. (p. 461)

4. A _____ is a customary unit for measuring liquid volume. (p. 457)

▶ **Concepts and Skills**

Complete.

5. 9 feet = _____ inches

6. 7 tons = _____ pounds

7. 10 pints = _____ cups

8. 4 decimeters = _____ centimeters

9. 8 liters = _____ milliliters

10. 5 weeks = _____ days

Compare using <, >, or =.

11. 3 yards ◯ 36 inches

12. 10 cups ◯ 80 fluid ounces

13. 4 pounds ◯ 96 ounces

14. 8 meters ◯ 700 centimeters

15. 6 liters ◯ 6,500 milliliters

16. 9 kilograms ◯ 9,000 grams

Add or subtract.

17. 8 hr 30 min
    − 6 hr 25 min

18. 7 c 4 fl oz
    + 4 c 3 fl oz

19. 9 yd 1 ft
    − 5 yd 2 ft

Fill in the bubble completely to show your answer.

20. Maya's band rehearsal started at 10:30 A.M. It ended 1 hour and 40 minutes later. At what time did Maya's band rehearsal end?

   Ⓐ 12:10 A.M.

   Ⓑ 8:50 A.M.

   Ⓒ 12:10 P.M.

   Ⓓ 11:10 P.M.

21. Darlene is making punch. She pours 4 quarts 2 cups of apple juice into a bowl. Then she pours 3 quarts 1 cup of grape juice into the bowl. How much juice is in the bowl now?

   Ⓐ 1 quart 1 cup

   Ⓑ 7 quarts 1 cup

   Ⓒ 7 quarts 3 cups

   Ⓓ 8 quarts 1 cup

22. Kainoa bought a brick of modeling clay that was labeled 2 kilograms. He needs to separate the clay into balls that are measured in grams. How many grams does he have?

   Ⓐ 20 grams

   Ⓑ 200 grams

   Ⓒ 2,000 grams

   Ⓓ 20,000 grams

23. A truck driver's truck weighs 3 tons. A weigh station measures the weight in pounds. How many pounds does the truck weigh?

   Ⓐ 600 pounds

   Ⓑ 2,000 pounds

   Ⓒ 3,000 pounds

   Ⓓ 6,000 pounds

Name _____

Fill in the bubble completely to show your answer.

24. Brody and Amanda canoed for 1 hour and 20 minutes before stopping to fish at 1:15 P.M. At what time did they start canoeing?

Ⓐ 11:55 A.M.

Ⓑ 12:05 P.M.

Ⓒ 2:35 P.M.

Ⓓ 11:55 P.M.

25. Lewis fills his thermos with 2 liters of water. Garret fills his thermos with 1 liter of water. How many more milliliters of water does Lewis have than Garret?

Ⓐ   1 more milliliter

Ⓑ   100 more milliliters

Ⓒ 1,000 more milliliters

Ⓓ 2,000 more milliliters

26. Lola won the 100-meter freestyle event at her swim meet. How many decimeters did Lola swim?

Ⓐ   1 decimeter

Ⓑ   10 decimeters

Ⓒ   100 decimeters

Ⓓ 1,000 decimeters

27. What is the best estimate for the length of an ant's leg?

Ⓐ 2 millimeters

Ⓑ 2 centimeters

Ⓒ 2 decimeters

Ⓓ 2 meters

► **Constructed Response**

28. Sabita made this table to relate two customary units of liquid volume. List the number pairs for the table. **Describe** the relationship between the numbers in each pair.

1	2
2	4
3	6
4	8
5	10

_____

_____

_____

29. Label the columns of the table. **Explain** your answer.

_____

_____

_____

► **Performance Task**

30. Landon borrowed a book from the library. The data show the lengths of time Landon read the book each day until he finished it.

Time Reading Book (in hours)
$\frac{1}{4}, \frac{1}{4}, 1, \frac{1}{4}, \frac{1}{2}, \frac{3}{4}, \frac{1}{2}, \frac{1}{4}$

A Make a tally table and a line plot to show the data.

Time Reading Book	
Time (in hours)	**Tally**

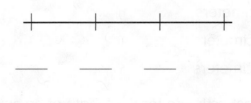

B **Explain** how you used the tally table to label the numbers and plot the Xs on the line plot.

_____

_____

_____

C What is the difference between the longest time and shortest time Landon spent reading the book? _____

Chapter 13

Algebra: Perimeter and Area

Show What You Know

Check your understanding of important skills.

Name _____

▶ **Missing Factors** Find the missing factor.

1.

_____ × 6 = 24

2.

3 × _____ = 27

▶ **Add Whole Numbers** Find the sum.

3. 17 + 153 + 67 = _____

4. 8 + 78 + 455 = _____

5. 211 + 52 + 129 + 48 = _____

6. 42 + 9 + 336 + 782 = _____

▶ **Multiply Whole Numbers** Find the product.

7. 78
  × 6

8. 29
  × 7

9. 42
  × 5

10. 57
  × 9

MATH DETECTIVE
WITH
CARMEN SANDIEGO™

Native Americans once lived near Cartersville, Georgia, in an area that is now a state park. They constructed burial mounds that often contained artifacts, such as beads, feathers, and copper ear ornaments. One of the park's mounds is 63 feet in height. Be a Math Detective. If the top of the mound is rectangular in shape with a perimeter of 322 yards, what could be the side lengths of the rectangle?

▶ **Visualize It** ••••••••••••••••••••••••••••••••••

Sort words with a ✓ using the Venn diagram.

Measurement

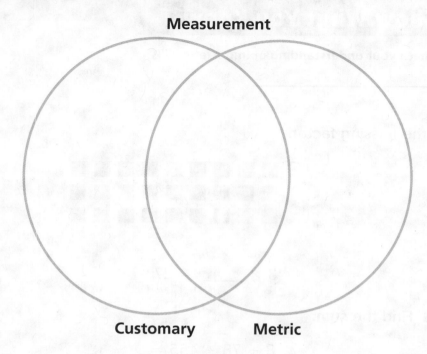

Customary   Metric

▶ **Understand Vocabulary** ••••••••••••••••••••••••••

Write the word or term that answers the riddle.

1. I am the number of square units needed to cover a surface.

   _____

2. I am the distance around a shape.

   _____

3. I am a unit of area that measures 1 unit by 1 unit.

   _____

4. I am a set of symbols that expresses a mathematical rule.

   _____

GO Online • eStudent Edition • Multimedia eGlossary

Name _____

Perimeter

Essential Question How can you use a formula to find the perimeter of a rectangle?

 UNLOCK the Problem REAL WORLD

Julio is putting a stone border around his rectangular garden. The length of the garden is 7 feet. The width of the garden is 5 feet. How many feet of stone border does Julio need?

Perimeter is the distance around a shape.

To find how many feet of stone border Julio needs, find the perimeter of the garden.

- Circle the numbers you will use.
- What are you asked to find?

_____

_____

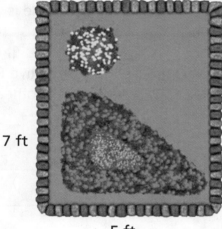

7 ft

5 ft

 Use addition.

Perimeter of a Rectangle = length + width + length + width

$\qquad$ 7 + 5 + 7 + 5 = _____

The perimeter is _____ feet.

So, Julio needs _____ feet of stone border.

 Use multiplication.

Ⓐ Find Perimeter of a Rectangle

Perimeter = (2 × length) + (2 × width)

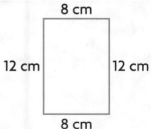

8 cm

12 cm   12 cm

8 cm

Perimeter = (2 × 12) + (2 × 8)

$\qquad$ = 24 + 16

$\qquad$ = _____

So, the perimeter is _____ centimeters.

Ⓑ Find Perimeter of a Square

Perimeter = 4 × one side

16 in.

16 in.   16 in.

16 in.

Perimeter = 4 × 16

$\qquad$ = _____

So, the perimeter is _____ inches.

MATHEMATICAL PRACTICES

Math Talk Explain how using addition and using multiplication to find the perimeter of a rectangle are related.

© Houghton Mifflin Harcourt Publishing Company

Use a Formula A **formula** is a mathematical rule. You can use a formula to find perimeter.

$$P = (2 \times l) + (2 \times w)$$

↑ perimeter   ↑ length   ↑ width

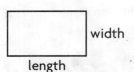
width
length

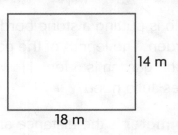

Example **Find the perimeter of the rectangle.**

$P = (2 \times l) + (2 \times w)$

$= (2 \times$ _____ $) + (2 \times$ _____ $)$ Think: Write the measures you know.

$=$ _____ $+$ _____   Think: Do what is in parentheses first.

$=$ _____

The perimeter of the rectangle is _____.

14 m

18 m

1. Can you use the Distributive Property to write the formula
   $P = (2 \times l) + (2 \times w)$ another way? **Explain.**

   _____

   _____

   _____

   _____

Try This! **Write a formula for the perimeter of a square.**

Use the letter _____ for perimeter.

Use the letter _____ for the length of a side.

Formula: _____

2. **Justify** the formula you wrote for the perimeter of a square.

   _____

   _____

   _____

   _____

Name _____

Share and Show

1. Find the perimeter of the rectangle.

$P = ($ _____ $\times$ _____ $) + ($ _____ $\times$ _____ $)$

$= ($ _____ $\times$ _____ $) + ($ _____ $\times$ _____ $)$

$=$ _____ $+$ _____

$=$ _____

The perimeter is _____ feet.

Formulas for Perimeter

Rectangle:
$P = (2 \times l) + (2 \times w)$ or
$P = 2 \times (l + w)$

Square:
$P = 4 \times s$

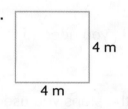
8 ft

4 ft

Find the perimeter of the rectangle or square.

2.
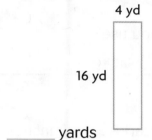
4 yd

16 yd

_____ yards

3.
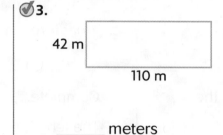
42 m

110 m

_____ meters

4.
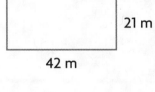
4 m

4 m

_____ meters

Math Talk MATHEMATICAL PRACTICES
Can you use the formula $P = (2 \times l) + (2 \times w)$ to find the perimeter of a square? Explain.

On Your Own

Find the perimeter of the rectangle or square.

5.
34 in.

20 in.

_____ inches

6.
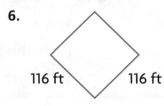
116 ft   116 ft

_____ feet

7.
21 m

42 m

_____ meters

8.
63 cm

42 cm

_____ centimeters

9.
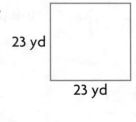
23 yd

23 yd

_____ yards

10.
40 in.

133 in.

_____ inches

🔑 UNLOCK the Problem REAL WORLD

11. Alejandra plans to sew fringe on a scarf. The scarf is shaped like a rectangle. The length of the scarf is 48 inches. The width is one half the length. How much fringe does Alejandra need?

Ⓐ 72 inches Ⓒ 120 inches

Ⓑ 96 inches Ⓓ 144 inches

a. Draw a picture of the scarf, and label the given measurements on your drawing.

b. What do you need to find?

_____

c. What formula will you use?

_____

d. Show the steps you use to solve the problem.

e. Complete.

The length of the scarf is _____ inches.

The width is one half the length

or _____ ÷ 2.

The width is _____ inches.

So, the perimeter is (_____ × _____) +

(_____ × _____) = _____ inches.

f. Fill in the bubble for the correct answer choice above.

---

12. 🔆 H.O.T. What is the side length of a square with a perimeter of 44 centimeters?

Ⓐ 4 centimeters

Ⓑ 11 centimeters

Ⓒ 22 centimeters

Ⓓ 176 centimeters

13. Mr. Wong is putting a brick edge around his rectangular patio. What is the perimeter of the patio?

18 ft

10 ft

Ⓐ 28 ft Ⓑ 38 ft Ⓒ 56 ft Ⓓ 66 ft

Name _____

Area

Essential Question How can you use a formula to find the area of a rectangle?

🔑 UNLOCK the Problem · REAL WORLD

The **base, b,** of a two-dimensional figure can be any side. The **height, h,** is the measure of a perpendicular line segment from the base to the top of the figure.

Remember

Perpendicular lines and perpendicular line segments form right angles.

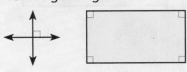

Area is the number of **square units** needed to cover a flat surface. A square unit is a square that is 1 unit long and 1 unit wide. To find the area of a figure, count the number of square units inside the figure.

How are the base, height, and area of a rectangle related?

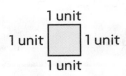

1 unit

1 unit ☐ 1 unit

1 unit

🔒 **Complete the table to find the area.**

Figure	Base	Height	Area
	5 units		

1. What relationship do you see among the base, height, and area?

_____

_____

2. Write a formula for the area of a rectangle. Use the letter A for area. Use the letter b for base. Use the letter h for height.

Formula: _____

MATHEMATICAL PRACTICES

Math Talk How do you decide which side of a rectangle to use as the base?

Use a Formula You can use a formula to find the area.

$$A = b \times h$$

↑ area   ↑ base   ↑ height

height

base

Examples Use a formula to find the area of a rectangle and a square.

A

6 ft

2 ft

$A = b \times h$

$= \underline{\hspace{1cm}} \times \underline{\hspace{1cm}}$

$= \underline{\hspace{1cm}}$

The area is _____.

B

2 m

2 m

$A = b \times h$

$= \underline{\hspace{1cm}} \times \underline{\hspace{1cm}}$

$= \underline{\hspace{1cm}}$

The area is _____.

Try This! Write a formula for the area of a square.

Use the letter _____ for area.

Use the letter _____ for the length of a side.

Formula: _____

Share and Show

1. Find the area of the rectangle.

$A = b \times \underline{\hspace{1cm}}$

$= \underline{\hspace{1cm}} \times \underline{\hspace{1cm}}$

$= \underline{\hspace{1cm}}$

11 cm

13 cm

Name 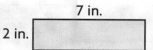 _____

Find the area of the rectangle or square.

2.
7 in.
2 in.

⊘ 3.

9 m   9 m

⊘ 4.
8 ft
14 ft

On Your Own

Find the area of the rectangle or square.

5.
13 ft
5 ft

6.
6 km
6 km

7.
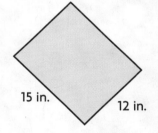
15 in.   12 in.

8.
29 m
10 m

9.
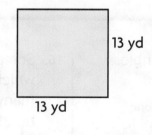
13 yd
13 yd

10.
2 cm
20 cm

Practice: Copy and Solve **Find the area of the rectangle.**

11. base: 16 feet
   height: 6 feet

12. base: 9 yards
   height: 17 yards

13. base: 14 centimeters
   height: 11 centimeters

🔑 UNLOCK the Problem REAL WORLD

14. Nancy and Luke are drawing plans for rectangular flower gardens. In Nancy's plan, the garden is 18 feet by 12 feet. In Luke's plan, the garden is 15 feet by 15 feet. Who drew the garden plan with the greater area? What is the area?

  Ⓐ Luke; 205 square feet   Ⓒ Nancy; 216 square feet

  Ⓑ Nancy; 206 square feet   Ⓓ Luke; 225 square feet

a. What do you need to find? _____

_____

b. What formula will you use? _____

c. What units will you use to write the answer? _____

d. Show the steps to solve the problem.

e. Complete the sentences.

The area of Nancy's garden is

_____.

The area of Luke's garden is

_____.

_____ garden has the greater area.

f. Fill in the bubble for the correct answer choice above.

---

15.  Find the area of the rectangle.

The length of one small square is 4 feet.

  Ⓐ   32 square feet

  Ⓑ   88 square feet

  Ⓒ   336 square feet

  Ⓓ   384 square feet

16. Sonia is buying carpet for the dining room, which measures 15 feet by 12 feet. How many square feet of carpet does Sonia need to cover the dining room?

  Ⓐ   45 square feet

  Ⓑ   54 square feet

  Ⓒ   170 square feet

  Ⓓ   180 square feet

Area of Combined Rectangles

Essential Question How can you find the area of combined rectangles?

🔑 UNLOCK the Problem · REAL WORLD

Jan is visiting a botanical garden with her family. The diagram shows two rectangular sections of the garden. What is the total area of the two sections?

There are different ways to find the area of combined rectangles.

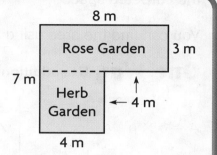

🔑 One Way Count square units.

Materials ▪ grid paper

• Draw the garden on grid paper. Then find the area of each section by counting squares inside the shape.

Rose Garden	Herb Garden
Area = _____ square meters	Area = _____ square meters

• Add the areas.

  _____ + _____ = _____ square meters

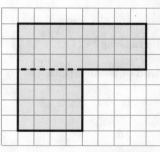

1 square = 1 square meter

🔑 Another Way Use the area formula for a rectangle.

Ⓐ Rose Garden

$A = b \times h$

= _____ × _____

= _____ square meters

Ⓑ Herb Garden

$A = b \times h$

= _____ × _____

= _____ square meters

• Add the areas.

  _____ + _____ = _____ square meters

So, the total area is _____ square meters.

Math Talk MATHEMATICAL PRACTICES
Is there another way you could divide the figure to find the total area? Explain.

🔒 Example

Greg is laying carpet in the space outside his laundry room. The diagram shows where the carpet will be installed. The space is made of combined rectangles. What is the area of the carpeted space?

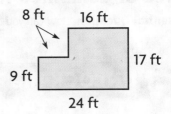

You can find the area using addition or subtraction.

One Way Use addition.

Rectangle A	Rectangle B
$A = b \times h$	$A = b \times h$
$= 8 \times \underline{\hspace{3em}}$	$= \underline{\hspace{3em}} \times 17$
$= \underline{\hspace{3em}}$	$= \underline{\hspace{3em}}$

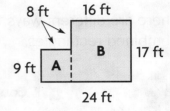

Sum of the areas:

$\underline{\hspace{6em}} + \underline{\hspace{6em}} = \underline{\hspace{6em}}$ square feet

Another Way Use subtraction.

Area of whole space	Area of missing section
$A = b \times h$	$A = b \times h$
$= 24 \times \underline{\hspace{3em}}$	$= \underline{\hspace{3em}} \times \underline{\hspace{3em}}$
$= \underline{\hspace{3em}}$	$= \underline{\hspace{3em}}$

Difference between the areas:

$\underline{\hspace{6em}} - \underline{\hspace{6em}} = \underline{\hspace{6em}}$ square feet

So, the area of the carpeted space is $\underline{\hspace{6em}}$ square feet.

- Is there another way you could divide the figure to find the total area? **Explain.**

$\underline{\hspace{30em}}$

$\underline{\hspace{30em}}$

$\underline{\hspace{30em}}$

$\underline{\hspace{30em}}$

Name _____

Share and Show

1. Explain how to find the total area of the figure.

_____

_____

_____

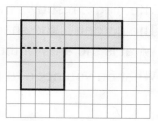

1 square = 1 square foot

Find the area of the combined rectangles.

2.
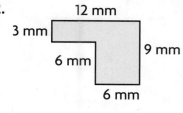

12 mm
3 mm
6 mm
9 mm
6 mm

_____

3.
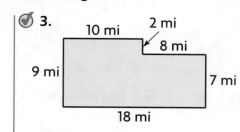

10 mi   2 mi
8 mi
9 mi   7 mi
18 mi

_____

4.
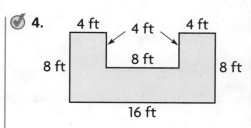

4 ft   4 ft   4 ft
8 ft   8 ft   8 ft
16 ft

_____

MATHEMATICAL PRACTICES

Math Talk Describe the characteristics of combined rectangles.

On Your Own

Find the area of the combined rectangles.

5.
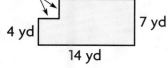

3 yd   11 yd
4 yd   7 yd
14 yd

_____

6.
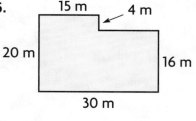

15 m   4 m
20 m   16 m
30 m

_____

7.
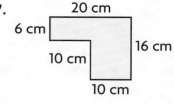

20 cm
6 cm
10 cm   16 cm
10 cm

_____

8. **H.O.T.** **Write Math** Explain how to find the perimeter and area of the combined rectangles at the right.

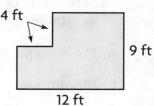

4 ft
9 ft
12 ft

_____

_____

_____

_____

UNLOCK the Problem   REAL WORLD

9. The diagram shows the layout of Mandy's garden. The garden is the shape of combined rectangles. What is the area of the garden?

a. What do you need to find?

_____

b. How can you divide the figure to help you find the total area?

_____

_____

c. What operations will you use to help you find the answer?

_____

d. Draw a diagram to show how you divided the figure. Then show the steps to solve the problem.

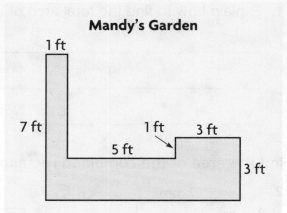

Mandy's Garden

So, the area of the garden is _____.

10. Test Prep Ms. Greene hired a contractor to remodel her house. What could the contractor use the area formula to find?

(A) the amount of fencing to put around the backyard

(B) the amount of floor the carpeting should cover

(C) the amount of water needed to fill the pool

(D) the amount of wallpaper border to put around the ceiling

FOR MORE PRACTICE:
Standards Practice Book, pp. P251–P252

Name _____

 Mid-Chapter Checkpoint

▶ **Vocabulary**

Choose the best term from the box.

1. A square that is 1 unit wide and 1 unit long is a

   _____. (p. 501)

2. The _____ of a two-dimensional figure can be any side. (p. 501)

3. A set of symbols that expresses a mathematical rule is

   called a _____. (p. 498)

4. The _____ is the distance around a shape. (p. 497)

▶ **Concepts and Skills**

Find the perimeter and area of the rectangle or square.

5.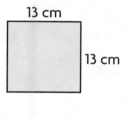

   13 cm
   13 cm

   _____

6. 21 ft
   3 ft

   _____

7.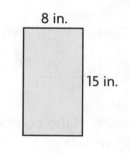

   8 in.
   15 in.

   _____

Find the area of the combined rectangles.

8.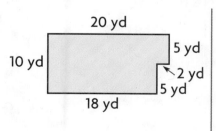

   20 yd
   10 yd
   5 yd
   2 yd
   5 yd
   18 yd

   _____

9.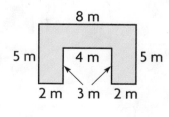

   8 m
   5 m   4 m   5 m
   2 m 3 m 2 m

   _____

10.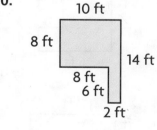

    10 ft
    8 ft
    14 ft
    8 ft
    6 ft
    2 ft

    _____

© Houghton Mifflin Harcourt Publishing Company

Fill in the bubble completely to show your answer.

11. Which figure has the greatest perimeter?

Ⓐ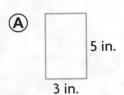
5 in.
3 in.

Ⓒ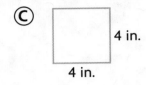
4 in.
4 in.

Ⓑ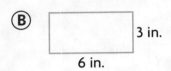
3 in.
6 in.

Ⓓ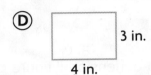
3 in.
4 in.

12. Which figure has an area of 108 square centimeters?

Ⓐ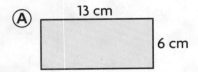
13 cm
6 cm

Ⓒ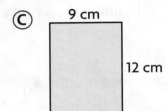
9 cm
12 cm

Ⓑ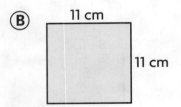
11 cm
11 cm

Ⓓ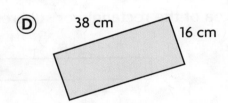
38 cm
16 cm

13. Which of the combined rectangles has an area of 40 square feet?

Ⓐ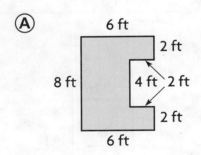
6 ft
2 ft
8 ft   4 ft 2 ft
2 ft
6 ft

Ⓒ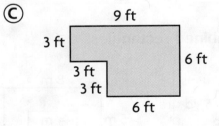
9 ft
3 ft
6 ft
3 ft
3 ft
6 ft

Ⓑ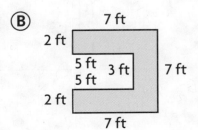
7 ft
2 ft
5 ft 3 ft 7 ft
5 ft
2 ft
7 ft

Ⓓ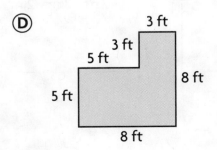
3 ft
3 ft
5 ft
8 ft
5 ft
8 ft

Name _____

Find Unknown Measures

Essential Question How can you find an unknown measure of a rectangle given its area or perimeter?

🔑 UNLOCK the Problem REAL WORLD

Tanisha is painting a mural that is in the shape of a rectangle. The mural covers an area of 54 square feet. The base of the mural measures 9 feet. What is its height?

Use a formula for area.

- **What do you need to find?**

  _____

- **What information do you know?**

  _____

🔑 Example 1 Find an unknown measure given the area.

MODEL

Think: Label the measures you know. Use n for the unknown.

$A =$ _____ $h =$ _____

$b =$ _____

So, the height of the mural is _____ feet.

RECORD

Use the model to write an equation and solve.

_____ = _____ _____   Write the formula for area.

_____ = _____ _____   Use the model to write an equation.

$54 = 9 \times$ _____   What times 9 equals 54?

The value of n is _____.

Think: n is the height of the mural.

Math Talk MATHEMATICAL PRACTICES
Explain how you can use division to find an unknown factor.

1. **What if** the mural were in the shape of a square with an area of 81 square feet? What would the height of the mural be? **Explain**.

_____

_____

2. **Explain** how you can find an unknown side length of any square, when given only the area of the square.

_____

_____

_____

🔑 Example 2 Find an unknown measure given the perimeter.

Gary is building an outdoor pen in the shape of a rectangle for his dog. He will use 24 meters of fencing. The pen will be 3 meters wide. How long will the pen be?

Use a formula for perimeter.

MODEL

Think: Label the measures you know. Use *n* for the unknown.

w = _____

l = _____

P = _____

RECORD

Use the model to write an equation and solve.

$P = (2 \times l) + (2 \times w)$

_____ = (_____ _____) + (_____ _____)

_____ = (_____ _____) + _____

Think: $(2 \times n)$ is an unknown addend.

24 = _____ + 6 **Think:** What is 24 − 6?

The value of $(2 \times n)$ is 18.

To find the value of *n*, find the unknown factor.

$2 \times$ _____ $= 18$

The value of *n* is _____.

Think: *n* is the length of the pen.

So, the pen will be _____ long.

Try This! The perimeter of a square is 24 feet. Find the side length.

Draw a model.	Write an equation.
	$P = 4 \times s$

Name _____

Share and Show

1. Find the unknown measure. The area of
   the rectangle is 36 square feet.

   $A = b \times h$

   _____ $= b \times$ _____

   The base of the rectangle is _____.

3 ft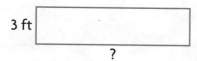

?

Find the unknown measure of the rectangle.

 2.

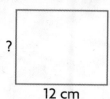

?

12 cm

Perimeter = 44 centimeters

width = _____

3.   9 in.

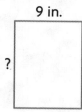

?

Area = 108 square inches

height = _____

 4.

5 m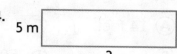

?

Area = 90 square meters

base = _____

On Your Own

Find the unknown measure of the rectangle.

5.

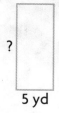

?

5 yd

Perimeter = 34 yards

length = _____

6.

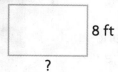

8 ft

?

Area = 96 square feet

base = _____

7.

?

9 cm

Area = 126 square centimeters

height = _____

Math Talk MATHEMATICAL PRACTICES
Explain how using the area formula helps you find the base of a rectangle when you know its area and height.

8. **H.O.T.** **Write Math** ▸ A square has an area of 49 square inches.
   Explain how to find the perimeter of the square.

   _____

   _____

   _____

   _____

Problem Solving > REAL WORLD

9. **H.O.T.** The area of a swimming pool is 120 square meters. The width of the pool is 8 meters. What is the length of the pool in centimeters?

_____

10. **Test Prep** An outdoor deck is 7 feet wide. The perimeter of the deck is 64 feet. What is the length of the deck?

  Ⓐ 14 feet   Ⓒ 39 feet

  Ⓑ 25 feet   Ⓓ 50 feet

Connect to Science

Mountain Lions

Mountain lions are also known as cougars, panthers, or pumas. Their range once was from coast to coast in North America and from Argentina to Alaska. Hunting and habitat destruction now restricts their range to mostly mountainous, unpopulated areas.

Mountain lions are solitary animals. A male's territory often overlaps two females' territories but never overlaps another male's. The average size of a male's territory is 108 square miles, but it may be smaller or larger depending on how plentiful food is. The average size of a female's territory is 54 square miles.

11. A male and female mountain lion have overlapped territories. The area of overlap is 28 square miles. Using the data above, how much of the male's and female's territory is not shared?

_____

_____

12. A male mountain lion has a rectangular territory with an area of 96 square miles. If his territory is 8 miles wide, what is the length of his territory?

_____

_____

Name _____

Problem Solving • Find the Area

Essential Question How can you use the strategy *solve a simpler problem* to solve area problems?

🔑 UNLOCK the Problem ⟩ REAL WORLD

A landscaper is laying turf for a rectangular playground. The turf will cover the whole playground except for a square sandbox. The diagram shows the playground and sandbox. How many square yards of turf will the landscaper use?

Use the graphic organizer below to solve the problem.

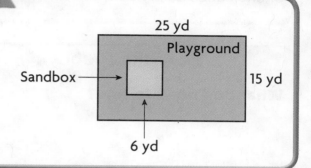

Read the Problem	Solve the Problem
What do I need to find? I need to find how many _____ the landscaper will use.	First, find the area of the playground. $A = b \times h$ $= \underline{\hspace{1cm}} \times \underline{\hspace{1cm}}$ $= \underline{\hspace{1cm}}$ square yards
What information do I need to use? The turf will cover the _____. The turf will not cover the _____. The length and width of the playground are _____ and _____. The side length of the square sandbox is _____.	Next, find the area of the sandbox. $A = s \times s$ $= \underline{\hspace{1cm}} \times \underline{\hspace{1cm}}$ $= \underline{\hspace{1cm}}$ square yards Last, subtract the area of the sandbox from the area of the playground. $\begin{array}{r} 375 \\ -\ 36 \\ \hline \end{array}$ square yards
How will I use the information? I can solve simpler problems. Find the area of the _____. Find the area of the _____. Then _____ the area of the _____ from the area of the _____.	So, the landscaper will use _____ _____ of turf to cover the playground. MATHEMATICAL PRACTICES **Math Talk** **Explain** how the strategy helped you to solve the problem.

Try Another Problem

Zach is laying a rectangular brick patio for a new museum. Brick will cover the whole patio except for a rectangular fountain, as shown in the diagram. How many square meters of brick does Zach need?

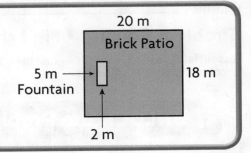

Read the Problem	Solve the Problem
What do I need to find?	
What information do I need to use?	
How will I use this information?	

- How many square meters of brick does Zach need? **Explain.**

_____

_____

_____

516

Name _____

Share and Show

UNLOCK the Problem **Tips**
√ Use the Problem Solving MathBoard.
√ Underline important facts.
√ Choose a strategy you know.

1. Lila is wallpapering one wall of her bedroom, as shown in the diagram. She will cover the whole wall except for the doorway. How many square feet of wallpaper does Lila need?

   First, find the area of the wall.

   $A = b \times h$

   = _____ × _____

   = _____ square feet

   Next, find the area of the door.

   $A = b \times h$

   = _____ × _____

   = _____ square feet

   Last, subtract the area of the door from the area of the wall.

   _____ − _____ = _____ square feet

   So, Lila needs _____ of wallpaper.

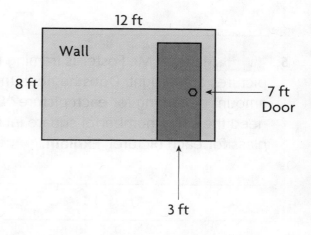

2. **H.O.T.** **What if** there was a square window on the wall with a side length of 2 feet? How much wallpaper would Lila need then? **Explain.**

   _____

   _____

   _____

3. Ed is building a model of a house with a flat roof, as shown in the diagram. There is a chimney through the roof. Ed will cover the roof with square tiles. If the area of each tile is 1 square inch, how many tiles will he need? **Explain.**

   _____

   _____

   _____

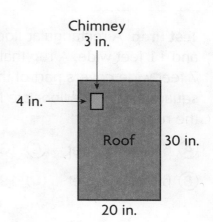

© Houghton Mifflin Harcourt Publishing Company

On Your Own................................

4. **H.O.T.** Lia has a dog and a cat. Together, the pets weigh 28 pounds. The dog weighs 3 times as much as the cat. How much does each pet weigh?

_____

5. **Write Math** ➤ Mr. Foster is framing the two pictures at the right. Does he need the same amount of framing for each picture? Does he need the same number of square inches of glass for each picture? **Explain.**

6 in.

5 in.

4 in.

5 in.

5 in.

SHOW YOUR WORK

_____

_____

_____

6. **H.O.T.** **What's the Error?** Claire says the area of a square with a side length of 100 centimeters is greater than the area of a square with a side length of 1 meter. Is she correct? **Explain.**

_____

_____

_____

_____

7. **Test Prep** A rectangular floor is 12 feet long and 11 feet wide. A rug that is 9 feet long and 7 feet wide covers part of the floor. How many square feet of the floor are NOT covered by the rug?

Ⓐ 63 square feet Ⓒ 132 square feet

Ⓑ 69 square feet Ⓓ 195 square feet

Chapter Review/Test

▶ Vocabulary

Choose the best term from the box.

1. The number of square units needed to cover a flat surface

   is the _____. (p. 501)

2. The distance around a shape is the _____. (p. 497)

▶ Concepts and Skills

Find the area of the rectangle or combined rectangles.

3.
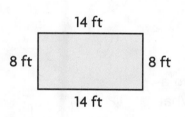

4.
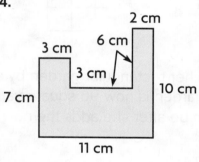

5.
16 yd
1 yd
6 yd   6 yd
4 yd   4 yd
8 yd   8 yd
12 yd

_____ | _____ | _____

Find the unknown measure of the rectangle.

6.
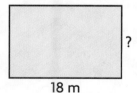
18 m

7.
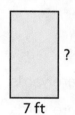
7 ft

8.
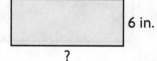
6 in.
?

Perimeter = 60 meters | Area = 91 square feet | Area = 60 square inches

width = _____ | height = _____ | base = _____

9. What is the perimeter of a rectangle with a length

   of 13 feet and a width of 9 feet? _____.

GO Online Assessment Options
Chapter Test

Fill in the bubble completely to show your answer.

10. Which pair of shapes has the same area?

Ⓐ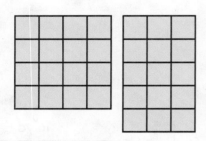

Ⓒ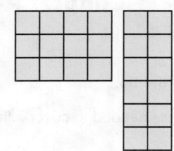

Ⓑ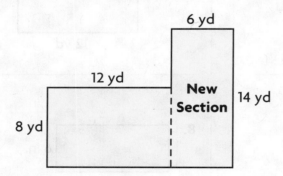

Ⓓ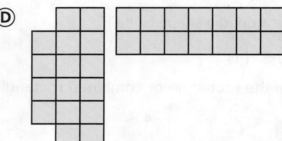

11. Jamie's mom wants to enlarge her rectangular garden by adding a new rectangular section. The garden is now 96 square yards. What will the total area of the garden be after she adds the new section?

6 yd

12 yd

New Section   14 yd

8 yd

Ⓐ 84 square yards   Ⓒ 180 square yards

Ⓑ 96 square yards   Ⓓ 192 square yards

12. A rectangular yoga studio has an area of 153 square feet. The width of the studio is 9 feet. What is the length of the studio?

Ⓐ 17 feet   Ⓒ 324 feet

Ⓑ 162 feet   Ⓓ 1,377 feet

Name _____

Fill in the bubble completely to show your answer.

13. Mr. Patterson had a rectangular deck with an area of 112 square feet built in his backyard. Which could be a diagram of Mr. Patterson's deck?

Ⓐ
10 ft

12 ft

Ⓒ
28 ft

4 ft

Ⓑ
13 ft

14 ft

Ⓓ
45 ft

11 ft

14. The town indoor pool is in a rectangular building. Marco is laying tile around the rectangular pool. How many square meters of tile will Marco need?

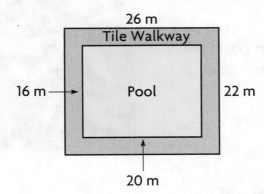

26 m
Tile Walkway

16 m → Pool 22 m

20 m

Ⓐ 96 square meters

Ⓑ 252 square meters

Ⓒ 572 square meters

Ⓓ 892 square meters

15. A drawing of a high school pool is shown below.

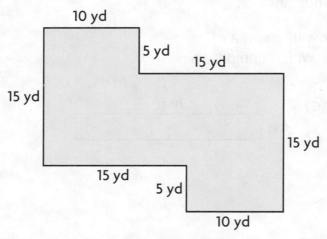

10 yd

5 yd

15 yd

15 yd

15 yd

15 yd

15 yd

5 yd

10 yd

What is the area of the pool? **Explain** how you know.

_____

_____

_____

► **Performance Task**

16. Mr. Brown has 24 meters of fencing. He wants to build a rectangular pen for his rabbits.

A Draw two different rectangles that Mr. Brown could build. Use only whole numbers for the lengths of the sides of each rectangle. Label the length of each side.

B Find the area in square meters of each rabbit pen you made in Part A. Show your work.

C If you were Mr. Brown, which of the two pens above would you construct for your rabbits? Explain why.

_____

_____

Table of Contents
Florida Lessons

Comparative Relational Thinking

Essential Question How can you determine if an equation is true without adding or subtracting?

🔑 UNLOCK the Problem

🔑 Example 1 Use comparative relational thinking to determine if an equation is *true* or *false*.

How can you tell if the following equation is true without adding?

$$36 + 47 = 40 + 43$$

Use comparative relational thinking to look and see how the addends on one side of the equation relate to those on the other side.

Think: 40 is _____ more than 36.

$$36 + 47 = 40 + 43$$

In order for the equation to be true, the second pair of addends must have a relationship that is equal but opposite.

Think: 43 must be _____ less than 47.

$$36 + 47 = 40 + 43$$

Since the relationships between the two pairs of addends undo each other, the equation is true.

🔑 Example 2

This method will also work for an equation involving subtraction. With a subtraction problem, you must add or subtract the same number to both number pairs on each side of the equation. Is the following equation true or false?

Think: 95 is _____ more than 86, so for

the equation to be true, 82 must be _____ more than 72.

$$86 - 72 = 95 - 82$$

Compare: 82 is _____ more than 72. Are the

relationships the same? _____

The relationships between the two pairs of numbers are different, so the equation is false.

Math Talk

MATHEMATICAL PRACTICES

Explain how you can change Example 2 so the equation is true.

© Houghton Mifflin Harcourt Publishing Company

Try This! Use comparative relational thinking to find the unknown number.

Erica and Jennifer each ran twice this week. Erica ran 63 minutes on the first day and 28 minutes on the second day. Jennifer ran the same total time as Erica. If Jennifer ran 32 minutes on the second day, how much time did she run the first day?

$63 + 28 = n + 32$

How is 32 related to 28?

32 is _____ more than 28.

So, n must be _____ fewer than 63.

$n = $ _____

So, Jennifer ran _____ minutes on the first day.

Share and Show

Use comparative relational thinking to tell whether the equation is *true* or *false*.

1. $42 + 15 = 39 + 18$

Think: 39 is _____ less than 42.
For the equation to be true,
18 must be _____ more than 15.
18 is _____ more that 15,
so the equation is _____.

2. $79 - 34 = 86 - 48$

Think: 86 is _____ more than 79.
For the equation to be true,
48 must be _____ more than 34.
48 is _____ more than 34,
so the equation is _____.

3. $58 + 35 = 56 + 38$

_____

4. $52 - 17 = 46 - 11$

_____

Use comparative relational thinking to find the unknown number.

5. $27 - 9 = n - 11$

$n = $ _____

6. $84 + 16 = n + 7$

$n = $ _____

7. $55 - 7 = 48 - n$

$n = $ _____

Name _____

On Your Own ....

Use comparative relational thinking to tell whether the equation is *true* or *false*.

8. $61 - 12 = 58 - 9$

_____

9. $23 + 38 = 43 + 19$

_____

10. $76 + 58 = 66 + 48$

_____

11. $24 - 13 = 32 - 21$

_____

12. $273 - 75 = 280 - 82$

_____

13. $125 + 98 = 120 + 93$

_____

Use comparative relational thinking to find the unknown number.

14. $67 - 32 = n - 41$

$n =$ _____

15. $23 + 19 = n + 15$

$n =$ _____

16. $53 - 35 = n - 41$

$n =$ _____

17. $71 - 42 = n - 52$

$n =$ _____

18. $102 + 71 = n + 75$

$n =$ _____

19. $112 + 94 = 115 + n$

$n =$ _____

20. **Write Math** ► On Monday, Renalta ran and walked for a total of 48 minutes. She walked 23 minutes and ran the rest. On Tuesday, she ran and walked 53 minutes. If she ran the same amount of time on both days, how much time did she walk on Tuesday? Use comparative relational thinking and an equation to **explain** your solution.

_____

_____

21. **H.O.T.** What can you say about the unknowns in the equation $x + 22 = y + 32$?

_____

_____

22. **Test Prep** Choose which value for n makes the equation below true.

$48 + 8 = n + 4$

(A) $n = 8$ (B) $n = 44$ (C) $n = 60$ (D) $n = 52$

H.O.T. What's the Error?

23. Ethan and Mandy both tried to solve the equation $14 + n = 16 + 17$ using comparative relational thinking. Ethan said $n = 15$. Mandy said $n = 33$. Neither student was correct. Describe their possible errors.

Ethan's Error	Mandy's Error
_____	_____
_____	_____
_____	_____
_____	_____
_____	_____
_____	_____
_____	_____

a. What is the correct solution to the equation? **Explain** your reasoning.

_____

_____

b. Could you find the unknown number by comparing 14 to the other addend 17, and n to 16? **Explain** why or why not.

_____

_____

_____

Glossary

© Houghton Mifflin Harcourt Publishing Company

Pronunciation Key

a	add, map	ē	equal, tree	m	move, seem	o͞o	pool, food	u̇	pull, book
ā	ace, rate	f	fit, half	n	nice, tin	p	pit, stop	û(r)	burn, term
â(r)	care, air	g	go, log	ng	ring, song	r	run, poor	yo͞o	fuse, few
ä	palm, father	h	hope, hate	o	odd, hot	s	see, pass	v	vain, eve
b	bat, rub	i	it, give	ō	open, so	sh	sure, rush	w	win, away
ch	check, catch	ī	ice, write	ô	order, jaw	t	talk, sit	y	yet, yearn
d	dog, rod	j	joy, ledge	oi	oil, boy	th	thin, both	z	zest, muse
e	end, pet	k	cool, take	ou	pout, now	<u>th</u>	this, bathe	zh	vision, pleasure
		l	look, rule	o͝o	took, full	u	up, done		

ə   the schwa, an unstressed vowel representing the sound spelled *a* in *above*, *e* in *sicken*, *i* in *possible*, *o* in *melon*, *u* in *circus*

Other symbols:
- • separates words into syllables
- ' indicates stress on a syllable

A

acute angle [ə•kyo͞ot′ ang′gəl] **ángulo agudo**
An angle that measures greater than 0° and less than 90° (p. 382)
Example:

acute triangle [ə•kyo͞ot′ trī′ang•gəl]
triángulo acutángulo A triangle with three acute angles (p. 386)
Example:

addend [a′dend] **sumando** A number that is added to another in an addition problem
Example: 2 + 4 = 6;
        2 and 4 are addends.

addition [ə•di′shən] **suma** The process of finding the total number of items when two or more groups of items are joined; the opposite operation of subtraction

A.M. [ā•em′] **a.m.** The times after midnight and before noon

analog clock [anəl• ôg kläk] **reloj analógico**
A tool for measuring time, in which hands move around a circle to show hours, minutes, and sometimes seconds
Example:

angle [ang′gəl] **ángulo** A shape formed by two line segments or rays that share the same endpoint (p. 382)
Example:

area [âr′ē•ə] **área** The number of square units needed to cover a flat surface (p. 501)
Example:

area = 9 square units

array [ə•rā′] **matriz** An arrangement of objects in rows and columns
Example:

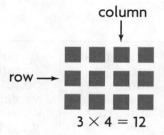

$$3 \times 4 = 12$$

Associative Property of Addition [ə•sō′shē•āt•iv präp′ər•tē əv ə•dish′ən] **propiedad asociativa de la suma** The property that states that you can group addends in different ways and still get the same sum
Example: $3 + (8 + 5) = (3 + 8) + 5$

Associative Property of Multiplication [ə•sō′shē•ə•tiv präp′ər•tē əv mul•tə•pli•kā′shən] **propiedad asociativa de la multiplicación** The property that states that you can group factors in different ways and still get the same product
Example: $3 \times (4 \times 2) = (3 \times 4) \times 2$

bar graph [bär graf] **gráfica de barras** A graph that uses bars to show data
Example:

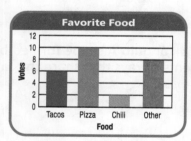

base [bās] **base** A polygon's side or a two-dimensional shape, usually a polygon or circle, by which a three-dimensional shape is measured or named (p. 501)
Examples:

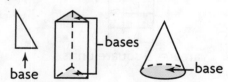

benchmark [bench′märk] **punto de referencia** A known size or amount that helps you understand a different size or amount (p. 250)

calendar [kal′ən•dər] **calendario** A table that shows the days, weeks, and months of a year

capacity [kə•pas′i•tē] **capacidad** The amount a container can hold when filled

Celsius (°C) [sel′sē•əs] **Celsius** A metric scale for measuring temperature

centimeter (cm) [sen′tə•mēt•ər] **centímetro (cm)** A metric unit for measuring length or distance 1 meter = 100 centimeters
Example:

1 centimeter

cent sign (¢) [sent sīn] **simbolo de centauo** A symbol that stands for *cent* or *cents*
Example: 53¢

clockwise [kläk′wīz] **en el sentido de las manecillas del reloj** In the same direction in which the hands of a clock move (p. 418)

closed shape [klōzd shāp] **figura cerrada** A two-dimensional shape that begins and ends at the same point
Examples:

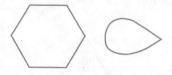

common denominator [käm′ən dē•näm′ə•nāt•ər] **común denominador** A common multiple of two or more denominators (p. 239)
Example: Some common denominators for $\frac{1}{4}$ and $\frac{5}{6}$ are 12, 24, and 36.

common factor [käm′ən fak′tər] **factor común** A number that is a factor of two or more numbers (p. 201)

common multiple [käm′ən mul′tə•pəl] **múltiplo común** A number that is a multiple of two or more numbers (p. 208)

Commutative Property of Addition
[kə•myōōt′ə•tiv präp′ər•tē əv ə•dish′ən] **propiedad conmutativa de la suma** The property that states that when the order of two addends is changed, the sum is the same
Example: 4 + 5 = 5 + 4

Commutative Property of Multiplication
[kə•myōōt′ə•tiv präp′ər•tē əv mul•tə•pli•kā′shən] **propiedad conmutativa de la multiplicación** The property that states that when the order of two factors is changed, the product is the same
Example: 4 × 5 = 5 × 4

compare [kəm•pâr′] **comparar** To describe whether numbers are equal to, less than, or greater than each other

compatible numbers [kəm•pat′ə•bəl num′bərz] **números compatibles** Numbers that are easy to compute mentally (p. 106)

composite number [kəm•päz′it num′bər] **número compuesto** A number having more than two factors (p. 211)
Example: 6 is a composite number, since its factors are 1, 2, 3, and 6.

corner [kôr′nər] **esquina** See *vertex*.

counterclockwise [kount•er•kläk′wīz] **en sentido contrario a las manecillas del reloj** In the opposite direction in which the hands of a clock move (p. 417)

counting number [kount′ing num′bər] **número natural** A whole number that can be used to count a set of objects (1, 2, 3, 4, . . .)

cube [kyōōb] **cubo** A three-dimensional shape with six square faces of the same size
Example:

cup (c) [kup] **taza (tz)** A customary unit used to measure capacity and liquid volume (p. 457)
1 cup = 8 ounces

data [dāt′ə] **datos** Information collected about people or things

decagon [dek′ə•gän] **decágono** A polygon with ten sides and ten angles

decimal [des′ə•məl] **decimal** A number with one or more digits to the right of the decimal point (p. 343)

decimal point [des′ə•məl point] **punto decimal** A symbol used to separate dollars from cents in money amounts, and to separate the ones and the tenths places in a decimal (p. 343)
Example: 6.4
                    ↑ decimal point

decimeter (dm) [des′i•mēt•ər] **decímetro (dm)** A metric unit for measuring length or distance (p. 467)
1 meter = 10 decimeters

degree (°) [di•grē′] **grado (°)** The unit used for measuring angles and temperatures (p. 421)

denominator [dē•näm′ə•nāt•ər] **denominador** The number below the bar in a fraction that tells how many equal parts are in the whole or in the group
Example: $\frac{3}{4}$ ← denominator

diagonal [dī•ag′ə•nəl] **diagonal** A line segment that connects two vertices of a polygon that are not next to each other
Example:

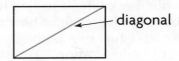

difference [dif′ər•əns] **diferencia** The answer to a subtraction problem

digit [dij′it] **dígito** Any one of the ten symbols 0, 1, 2, 3, 4, 5, 6, 7, 8, or 9 used to write numbers

digital clock [dij′i•təl kläk] **reloj digital** A clock that shows time to the minute, using digits
Example:

dime [dīm] **moneda de 10¢** A coin worth 10 cents and with a value equal to that of 10 pennies; 10¢
Example:

dimension [də•men′shən] **dimensión** A measure in one direction

Distributive Property [di•strib′yoo•tiv präp′ər•tē] **propiedad distributiva** The property that states that multiplying a sum by a number is the same as multiplying each addend by the number and then adding the products (p. 61)
Example: $5 \times (10 + 6) = (5 \times 10) + (5 \times 6)$

divide [də•vīd′] **dividir** To separate into equal groups; the opposite operation of multiplication

dividend [dəv′ə•dend] **dividendo** The number that is to be divided in a division problem
Example: $36 \div 6$; $6\overline{)36}$; the dividend is 36.

divisible [də•viz′ə•bəl] **divisible** A number is divisible by another number if the quotient is a counting number and the remainder is zero (p. 198)
Example: 18 is divisible by 3.

division [də•vi′zhən] **división** The process of sharing a number of items to find how many equal groups can be made or how many items will be in each equal group; the opposite operation of multiplication

divisor [də•vī′zər] **divisor** The number that divides the dividend
Example: $15 \div 3$; $3\overline{)15}$; the divisor is 3.

dollar [däl′ər] **dólar** Paper money worth 100 cents and equal to 100 pennies; $1.00
Example:

elapsed time [ē•lapst′ tīm] **tiempo transcurrido** The time that passes from the start of an activity to the end of that activity

endpoint [end′point] **extremo** The point at either end of a line segment or the starting point of a ray

equal groups [ē′kwəl groopz] **grupos iguales** Groups that have the same number of objects

equal parts [ē′kwəl pärts] **partes iguales** Parts that are exactly the same size

equal sign (=) [ē′kwəl sīn] **signo de igualdad** A symbol used to show that two numbers have the same value
Example: $384 = 384$

equal to [ē′kwəl too] **igual a** Having the same value
Example: $4 + 4$ is equal to $3 + 5$.

equation [ē•kwā′zhən] **ecuación** A number sentence which shows that two quantities are equal
Example: $4 + 5 = 9$

equivalent [ē•kwiv′ə•lənt] **equivalente** Having the same value or naming the same amount

equivalent decimals [ē•kwiv′ə•lənt des′ə•məlz] **decimales equivalentes** Two or more decimals that name the same amount (p. 352)

equivalent fractions [ē•kwiv′ə•lənt frak′shənz] **fracciones equivalentes** Two or more fractions that name the same amount (p. 227)
Example: $\frac{3}{4}$ and $\frac{6}{8}$ name the same amount.

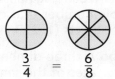

$$\frac{3}{4} = \frac{6}{8}$$

estimate [es′tə•māt] *verb* **estimar** To find an answer that is close to the exact amount

estimate [es′tə•mit] *noun* **estimación** A number that is close to the exact amount (p. 17)

even [ē′vən] **par** A whole number that has a 0, 2, 4, 6, or 8 in the ones place

expanded form [ek•span′did fôrm] **forma desarrollada** A way to write numbers by showing the value of each digit (p. 9)
Example: $253 = 200 + 50 + 3$